The Dual-Heartland Geography of the Book of Mormon

The Dual-Heartland Geography of the Book of Mormon

Edwin Goble

TEMPLE HILL BOOKS

ISBN 978-1-4341-0586-8

Copyright © 2025 Edwin Goble. All rights reserved.

Published by Temple Hill Books, an imprint of the Editorium

Temple Hill Books™ and Editorium™ are trademarks of:

The Editorium, LLC
West Jordan, UT 84081-6132
www.editorium.com

The views expressed in this book are the responsibility of the author and do not necessarily represent the position of Temple Hill Books. The reader alone is responsible for the use of any ideas or information provided by this book.

Contents

	Introduction: My Journey to the Dual-Heartland Model	vii
1	Understanding Book of Mormon Geography: Definitions, Methodologies, Models and Maps	1
2	Lehi's Landing: Traditions and Possibilities	19
3	Joseph Smith's Geographical Insights and the Hopewell Connection	27
4	The Zelph Incident: A Key to the Land Northward	41
5	Narrow Neck of Land and Narrow Pass: The Case for Chivela Pass	48
6	The Olmec as the Jaredites: San Lorenzo and the Narrow Neck	56
7	The River Sidon: Usumacinta vs. Grijalva—Usumacinta Wins	62
8	Cumorah in New York: Internal Geography and Archaeological Expectations	72
9	Other Early Migrations and Ancient Connections	91
10	Indigenous Legends and the Destruction of the "White People"	95

11	Internal Textual Evidence for a Limited, Tropical Geography, at Least for the Land Southward	147
12	John L. Sorenson's Foundational Mesoamerican Model	152
13	Conclusion: Converging Arguments for a Mesoamerican Land Southward and a North American Land Northward	154

Introduction: My Journey to the Dual-Heartland Model

My journey in understanding Book of Mormon geography has been one of continuous re-evaluation and refinement. From my early childhood in the 1970s, growing up in my grandparents' home filled with Latter-day Saint books, and discussions with family members on topics like this, I developed a strong belief in the Hemispherical Model of Book of Mormon geography. This pervasive view posited that Book of Mormon lands spanned vast portions of North and South America, with a particular emphasis on the Hill Cumorah in New York as the site of the great Nephite and Jaredite destructions, not merely the place where Moroni buried the plates. I devoured these classic Latter-day Saint texts and absorbed the points of view they presented, taking them very seriously. My grandfather's study room, filled with bookshelves addressing various gospel subjects, including early Book of Mormon archaeology, deeply influenced me and instilled a belief in this expansive geography. At that time, few doubted that Lehi landed in Chile, Panama was the Narrow Neck, and Zarahemla was somewhere in Colombia.

However, my scientific inclinations gradually clashed with some of these deeply ingrained geographical assumptions. A significant cognitive dissonance arose from statements made by early Church leaders regarding specific Book of Mormon locations within the United States. For instance, **Joseph Fielding Smith** in his *Doctrines of Salvation* quoted accounts identifying an "ancient site of the City of Manti" in Randolph County, Missouri, near Huntsville. While Smith presented these quotes in support of the traditional Hemispherical Model, I found them to be a poor fit with what I knew of the Book of Mormon's internal geography. Manti was consistently placed in the Land Southward, far from Missouri. This contradiction confused me throughout the time my mission, especially, when I was continually meditating upon such things, and planted seeds of determination to resolve the issue. I constantly sought for a rational framework in the real world that would make sense of Book of Mormon geography alongside a New York Cumorah placement, seeking to find a sensible, mostly skeletal, framework in the real world for the rest of a geography model that would accompany the Cumorah placement in New York, of where the Nephites were destroyed.

After my mission, armed with Sorenson's *The Geography of Book of Mormon Events: A Source Book*, I was ready to critically reassess the Hemispherical Model. It was around this time that I wrestled with my deeply held conviction about the New York Cumorah. My rational mind at times attempted to accept the "Two Cumorah" theory—the idea that all Book of Mormon events occurred in Mesoamerica, with a different Cumorah there—but I found I couldn't deny it without discomfort. A deep seated, soul-level connection to what I have always viewed as the truth on that placement always brought me back to accepting the New York Cumorah as a fact, that it was indeed not just where Moroni buried the plates, but also, that it was the place of the Nephite and Jaredite destructions. Thus, any viable geography model for me had to include the New York Cumorah.

The challenge then became how to reconcile a New York Cumorah with a limited geography (following the Mesoamericanist fixation on limited geography theories (i.e. the idea that Cumorah was not very far from the rest of the lands of the Book of Mormon), because of their beliefs about internal evidences within the Book of Mormon text for this. Trying to come up with some kind of theory like that was especially difficult in light of the perplexing Manti in Missouri statements, where the claim was made that a city of Manti was in the vicinity of Huntsville, Missouri. Because, at the time, it became clear to me that Huntsville, Missouri is not many thousands of miles from the Palmyra, New York, but only a little less than one thousand. And this is why the Missouri Manti issue became very attractive to me, and at the time, I believed that the information in those statements must have come from Joseph Smith. I sought to find a way to incorporate it into some kind of theory that also included a New York Cumorah, almost obsessively at the time. I struggled to find a convincing "Narrow Neck of Land" in the United States that could logically connect these points, initially assuming it had to be an hourglass-shaped isthmus surrounded by water on both sides, a concept influenced by Mesoamericanist rhetoric and Tradition Church about early hemispherical theories that included Panama as the candidate for the narrow neck of land spoken of in the Book of Mormon. For a time, I even entertained the unlikely geological idea that parts of North America might have been underwater at some point, forming such a neck between New York and Missouri. But quickly afterward, I reasoned that such a thing would have to show up in the geologic record, leading to its quick dismissal in my mind.

A significant epiphany came from reading material by FARMS scholars, particularly a review by John Clark of Richard Hauck's book, *Deciphering the Geography of the Book of Mormon*. Clark's review is called "A Key for Evaluating Nephite Geographies." Hauck proposed that the scriptural descriptions of the Narrow Neck could be interpreted to support a "coastal corridor" in Mesoamerica. Suddenly, I realized that the "Narrow Neck of Land" didn't *have* to be a traditional isthmus; this opened up possibilities for applying similar reasoning to North America. This new way of thinking, akin to Jonathan Neville's "Multiple Working Hypotheses" approach, helped me break free from traditional constraints. For a period, I even modified Hauck's idea, applying it to the area of the United States, to propose a coastal corridor along Lake Erie's southern shore as a candidate for the narrow neck, as this kind of thinking opened up a door wide-open for hypothetical North American placements.

Further seeming breakthroughs at the time came from a personal encounter with Duane Erickson at a conference, whose theory I had previously dismissed as far-fetched. Erickson proposed the Mississippi River as the Sidon and a Zarahemla candidate across from Nauvoo in Iowa. While initially skeptical, I was persuaded when he quoted Doctrine and Covenants 125:3, which names a modern city to be built as "Zarahemla" in that very area. Erickson's belief on this scripture has been carried over into virtually all modern variants of the Heartland theory. This correlation, combined with the proximity to the Missouri Manti, made his ideas seem more plausible for a U.S. setting. I reasoned that the Missouri River could be a "head" or tributary of the Mississippi (Sidon), even if the Mississippi itself didn't flow due north, in order to accommodate the Missouri Manti with Erickson's beliefs about Zarahemla and the Sidon. This led me to a temporary embracement of a U.S.-centric Heartland theory. I was careful to credit Erickson for these elements, as well as Duane Aston and Delbert Curtis for their contributions to my thinking, such as Aston's compelling Native American place-name evidence that "Niagara" means "neck of land." This linguistic parallel, combined with the Book of Mormon's description of the Narrow Neck as where "the sea divides the land," led me to initially consider the Niagara Peninsula as the Narrow Neck and the Great Lakes as the "West Sea." This period of exploration, while eventually refined, was essential for developing a framework that could bridge the distances required by the New York Cumorah.

This period also saw my unfortunate, albeit necessary, collaboration with Wayne May. The modern Heartland Models would not be what they are today without my historic teaming up with May in

Introduction: My Journey to the Dual-Heartland Model   xi

that time, and the Heartland Theory in general has long outlived my involvement with it. I believe that it was the Lord's plan that some form of the Heartland Theory would emerge eventually as a popular theory in the Church with or without my involvement in it. I don't say this because I believe the Heartland Theory is correct. I say it because I believe that the Lord allowed an alternate point of view to the Mesoamerican theory to emerge, so that the two would "run together"[1] in the marketplace of ideas. Ultimately, I believe this was so in order that that interaction between those two models would eventually give birth to the modern "Continental-style" models like Andrew Hedges' model, my Dual-Heartland model and Lance Weaver's "Continental Highland" model. Permit me to explain how this all came to be.

In the late 1990s and early 2000's, I partnered with Wayne May, who publishes *Ancient American Magazine*. At times, I wavered back and forth whether I would follow through with publishing a book that we were partnering on. The primary point of contention was May's insistence on including questionable artifacts, like the Soper-Savage and Burrows Cave relics, which mainstream archaeologists (and I) doubted. Then came incidents in the fall of 2001 that drove me to a point of desperation to publish my nascent manuscript, where I was facing severe financial hardship after the 9/11 attacks and my father's death.

Despite my strong misgivings, I felt compelled by my desperate circumstances to compromise, agreeing to a "neutral" position in the book that we would both take in print, calling for scientific testing of these artifacts. However, this proved disastrous. Scholars associated with FAIR (Foundation for Apologetic Information and Research), particularly Brant Gardner, publicly revealed that these artifacts had *already* been tested and found fraudulent *before* our book was published. Gardner then publicly released my private email retracting our neutral stance we had taken in the book on

---

1. 2 Nephi 29:8

those artifacts, causing my relationship with Wayne May to end and leading to my signing over the book's rights to him. This experience taught me a profound lesson about avoiding compromise on core principles, even in the face of extreme personal difficulty, especially regarding questionable archaeological evidence (whether one takes a "neutral" position or not), a lesson that shaped my subsequent methodological approach.

Nevertheless, this whole episode led to the popularization of the basic paradigm of the Heartland model, following the approach of the initial version that appeared in Wayne's (and my) book. It led to a number of new versions of the Heartland model to appear in subsequent years, despite my entire abandonment of the Heartland model.

This challenging episode, however, ultimately propelled me toward my current, more refined geographical model: the **Dual-Heartland Model**, also known as the **Semi-Limited Mesoamerican Theory**. This theory posits that the Book of Mormon's Land Southward, including Zarahemla and Nephi, is located in Mesoamerica, while the Land Northward, encompassing the Hopewell cultural sphere, extends into the Midwestern United States, culminating at the Hill Cumorah in New York. This view, lying within the spectrum between a full hemispherical and a purely limited Mesoamerican model, asserts that most of the Book of Mormon narrative takes place in Mesoamerica, with migrations and later historical events extending into the northern areas. And these northward migrations mixing in with inhabitants already in the area led to a relatively large population in the populous urban centers in Illinois and Ohio. Therefore, there is one urban "heartland" in Mesoamerica, with a satellite urban center (though not with the same amount of population) to the north (therefore being called the "dual-heartland" model). An analogy to this might be the fact that the Salt Lake-Provo-Ogden Metro Area in Utah is indeed "urban" with a significant population, as are many other scattered urban centers across the United States. Yet, the degree of urbanization and the amount of population in these scattered centers in each case is to a smaller degree than either that of California or on the Eastern Seaboard. The idea that this type of continental-spanning geography (or semi-limited or northern-continental) may have been in the minds of Joseph Smith and early Saints is further supported by scholarly analysis, as Andrew Hedges has shown.

My research continues to be driven by a belief that true understanding comes from harmonizing all available information, including revealed knowledge and academic findings. Truth can emerge from unexpected places, even from diverse and seemingly contradictory testimonies, perhaps as "portions of truth" that can eventually unite into one great whole.[2] Ultimately, the intellectual pursuit of truth should not be hindered by contention or personal biases, but guided by a spirit of inquiry and respect, always remembering that the ultimate purpose is to build faith and bear witness of the Lord and His word. My interactions with scholars from various backgrounds have reinforced the importance of focusing on shared humanity and mutual respect over intellectual disagreements, recognizing that true compassion transcends differing viewpoints.

---

2. 2 Nephi 29:7-11

Chapter 1

Understanding Book of Mormon Geography: Definitions, Methodologies, Models and Maps

Defining the geographical scope of the Book of Mormon is paramount to any serious study of its historicity. The text itself does not provide modern coordinates, leading to a complex and often contentious debate among scholars and enthusiasts. There can be no genuine Book of Mormon archaeology until a plausible real-world setting is identified and grounded in rigorous methodology. It is insufficient to merely unearth any ancient artifact and arbitrarily label it a "Book of Mormon site." To distinguish between a Book of Mormon artifact and any other ancient American artifact, it is necessary to know you are digging in the right place, and it is necessary to date the artifacts scientifically to know they are from the appropriate time period. This means that acceptable archaeological evidence must be scientifically dated under controlled conditions.

Book of Mormon Geographers of the Mesoamericanist variety usually come up with what they call an "internal" or "conceptual" map to visualize Book of Mormon Lands. Usually, these Mesoamericanists that envision the Book of Mormon Lands think of an Hourglass shape with the Narrow Neck of Land being flanked by two seas, because of the assumption that the Narrow Neck of Land is an isthmus. The following is an example of this kind of map:

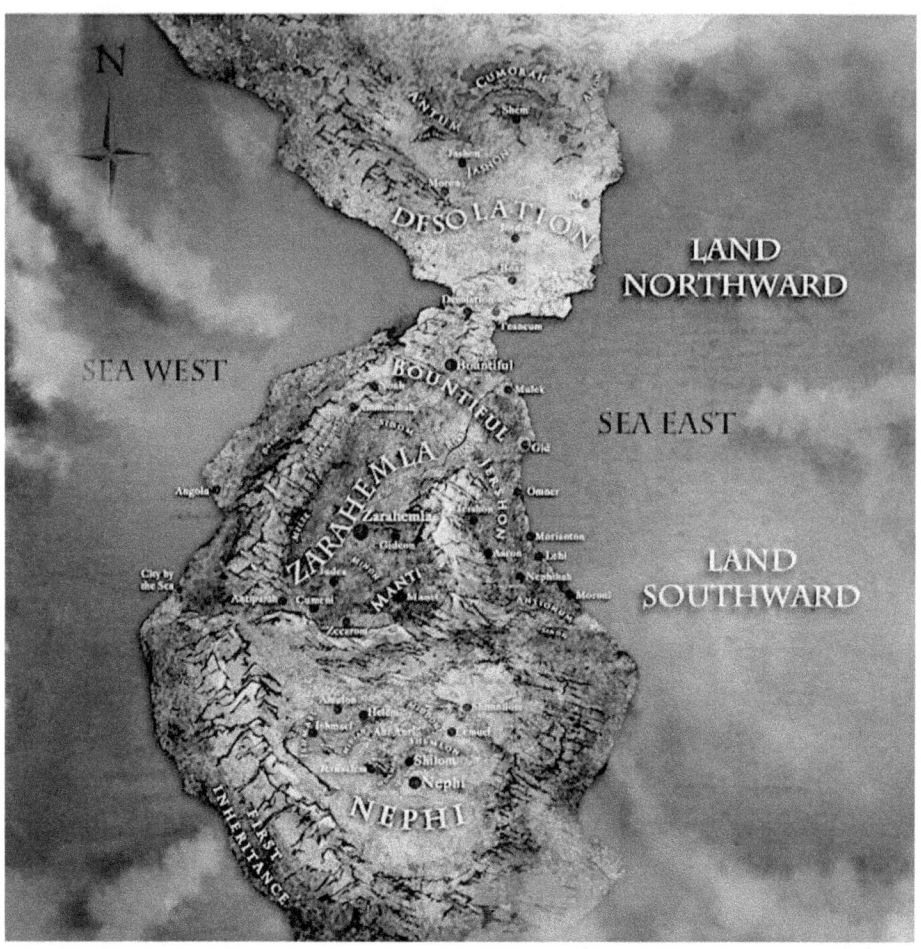

As you can see, Book of Mormon Geographers that make maps like these have already sabotaged themselves using the assumption of the isthmus configuration with the hourglass shape. Then they proceed to find something like it in the real world for their proposal. I don't think this is useful, because it cements that one flawed expectation into our minds, and sabotages us from the start. What is better as a first step is to understand the overall relational configuration of lands with that most significant landmark of all in the text, what is called the **Narrow Neck of Land**. So, instead of trying to define what a Narrow Neck of Land is at this point, instead I will just have it as a feature in the land in relation to the other major lands. This is also going to be a conceptual map of sorts too, but without an assumption that the Narrow Neck of Land is what people think it is.

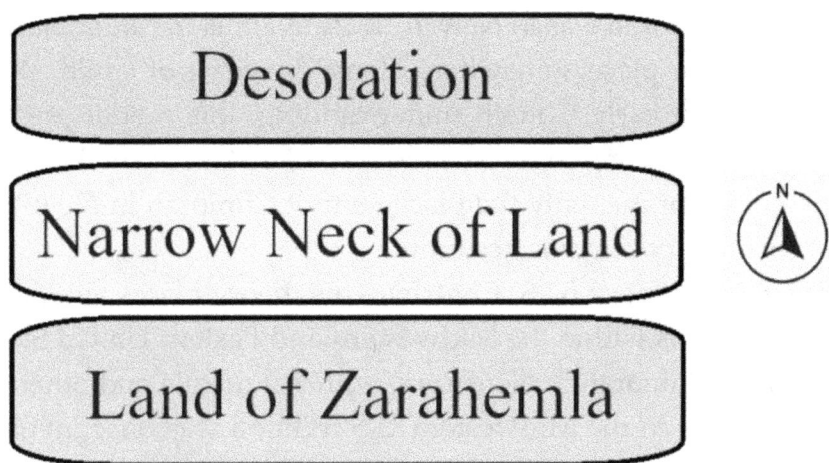

Now, this has shown us that Desolation as the Land Northward is North, the Narrow Neck of Land in general is South of that, and the general Land of Zarahemla is, in general, South of the Narrow Neck.

Key Geographical Models and Key Features with Maps in the New Proposal

Let us begin by defining the primary geographical models that have shaped this discussion (this is not an exhaustive list):

Mesoamerica: This is a technical archaeological term referring to a cultural area spanning approximately from central Mexico through Belize, Guatemala, El Salvador, Honduras, Nicaragua, and Costa Rica. It was the cradle of numerous pre-Columbian societies that flourished before the 16th-century Spanish colonization. Within this region, a popular theory today is the "Limited Mesoamerican theory," which places *all* Book of Mormon events, including Cumorah, within this relatively confined area. This is not my belief.

Hemispherical Theory: This model posits that Book of Mormon territory covered the entirety of North and South America, with the Isthmus of Panama serving as the Narrow Neck of Land. In this view, Cumorah was in New York, Zarahemla in Colombia, and Lehi's landing place somewhere along the coast of Chile. While widespread in early Church understanding, this model presents significant challenges in reconciling textual distances and historical plausibility. For me, only the placement of Cumorah in New York from this theory remains accurate.

Heartland Theory (U.S. Centric): This theory places all Book of Mormon events within the Midwestern and Eastern United States. It positions Cumorah in New York, with Zarahemla and other key Nephite lands in the Midwestern U.S. While I was once involved with this theory, even co-writing a book about it, my understanding has evolved significantly over two decades. Most iterations or variants of this model specifically propose a Manti candidate in Missouri and a Zarahemla candidate in Iowa.

Dual-Heartland Model (Semi-Limited Mesoamerican Theory): This is my current proposal. It maintains Cumorah in New York and identifies a "Northern Nephite domain" in the Midwestern U.S., specifically correlating with the Hopewell-Adena culture, the northern "heartland." However, it places the primary Book of Mormon narrative in terms of the time before the northward migrations, to include the Land Southward (Zarahemla and Nephi) and the Narrow Neck of Land, in Mesoamerica. This model is neither fully hemispherical nor purely limited to Mesoamerica; it spans a continental distance from Mesoamerica to New York, recognizing two distinct but connected heartlands of Book of Mormon civilization. The internal evidences of the Book of Mormon, when carefully studied, suggests a limited geography **in the first part of their history**, much less than a continent-spanning civilization, which makes a range of 800–1000 miles plausible, rather than thousands of miles. **This is indeed very limited**. It is only later, **after both Book**

of Mormon civilizations (both the Jaredite and the Lehite) started northward migrations that their domain ended up encompassing a span of this larger distance from Mesoamerica to New York. This is part of the reason that the Mesoamericanists have failed to a degree in their deductive reasoning on this point. This is much like the "Manifest Destiny" of the United States, starting out in a limited area in the east with the first colonies, and then spreading across the whole Continent to the west, with the exception that the Nephite and Jaredite expansions were from South to North.

The following map shows the main idea of this theory, with Cumorah and an urban Northern Nephite Domain anciently in the Eastern United States, and that region was part of the larger Land Northward or Land of Desolation. This Land of Desolation extended from Cumorah down to the Tehuantepec Isthmus in Mexico. Then the general Land of Zarahemla was in Southern Mexico and Guatemala, an area known as Mesoamerica:

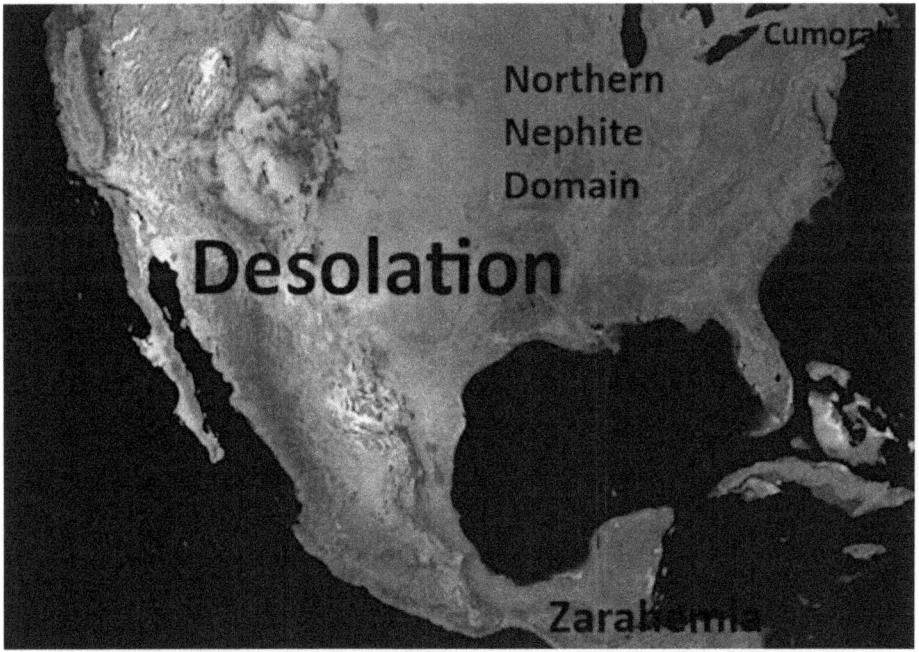

This map shows things from a continental view. Now, we will focus in with a more close-up view to the area of Mesoamerica:

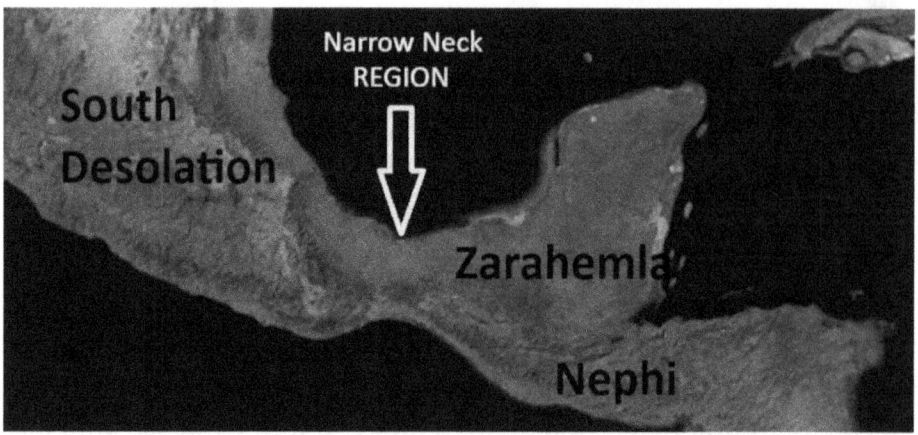

From this closer look, you can see how the southern part of Desolation extends left from the **Isthmus of Tehuantepec**. The Land of **Zarahemla** was generally in the lowland areas just east of the isthmus, while the Land of **Nephi** was in the highlands to the south. A yellow arrow on the map points to the Isthmus of Tehuantepec, which I now refer to as the **Narrow Neck Region**. I am using this language for an important reason instead of calling Tehuantepec the Neck of Land itself. This is because, in the Book of Mormon, the Jaredite heartland is close to the neck, and the Olmec heartland is in this region, right on the Gulf of Mexico side of the Isthmus. Historically, many theories about the location of the Book of Mormon lands have identified the Isthmus of Tehuantepec as the **Narrow Neck of Land**. I also made this assumption in my earlier book, *Resurrecting Cumorah*, some of which has been revised and included in this book.

However, I no longer believe the Tehuantepec Isthmus is the Narrow Neck of Land. I've wrestled with that idea for a long time, finally settling on the fact that the isthmus is just too fat to be considered narrow. I realized that the neck has to be a separate feature of the land in that region. Nevertheless, I still believe it lies within the *region* of the neck. And so, instead, I have gone back, yet again to adopt and modify a concept from F. Richard Hauck's 1980s book, *Deciphering the Geography of the Book of Mormon*. As you read above, this concept has attracted me before in past theories that I tried to construct. And, yet again, it becomes useful and key, by liberating us from the necessity to assume that the Neck of Land is either a peninsula or an isthmus. Hauck proposed a **Coastal Corridor** along the Pacific side of Tehuantepec. I've built on this idea again, thinking that his corridor does have something to do with the neck of land, because it does lead up to the point where the Sierra Madre mountain range splits into its western and eastern parts.

I propose, however, that the **Narrow Pass or Passage** mentioned in the Book of Mormon refers specifically to the **Chivela Pass**, which is this place where the Sierra Madre Mountains divide, not specifically to Hauck's corridor. But, Hauck's insight that this "pass" didn't need to be flanked by seas on both sides resonated with me. And now, I believe that the terms "Narrow Neck" and "Narrow Pass" describe the same geographic feature. While we often think of a "neck" as a landform with water on either side (an isthmus), the term can also describe any confined area hemmed in by geographical barriers, such as mountain ranges. Considering this, it's clear to me now that the **Chivela Pass itself is the Narrow Neck of Land or Small Neck of Land.** You will be able to see in general where this pass is on the following map:

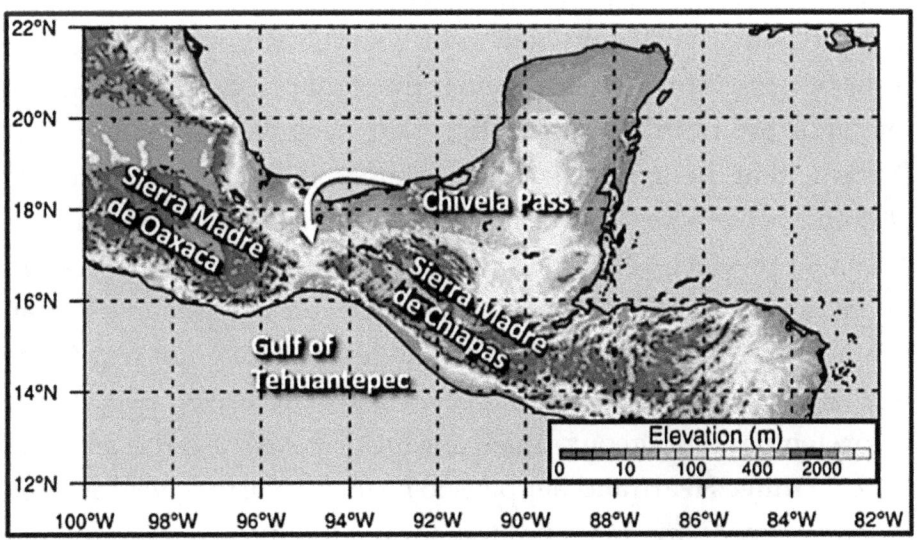

Where the brown is on this map is where the two halves of the Sierra Madre Mountains are, and you can see the arrow pointing to where the **Chivela Pass** is at, in the split in the ranges. Let's zoom in even closer to view precisely where this is on the following map. The red pointer will show us where it is:

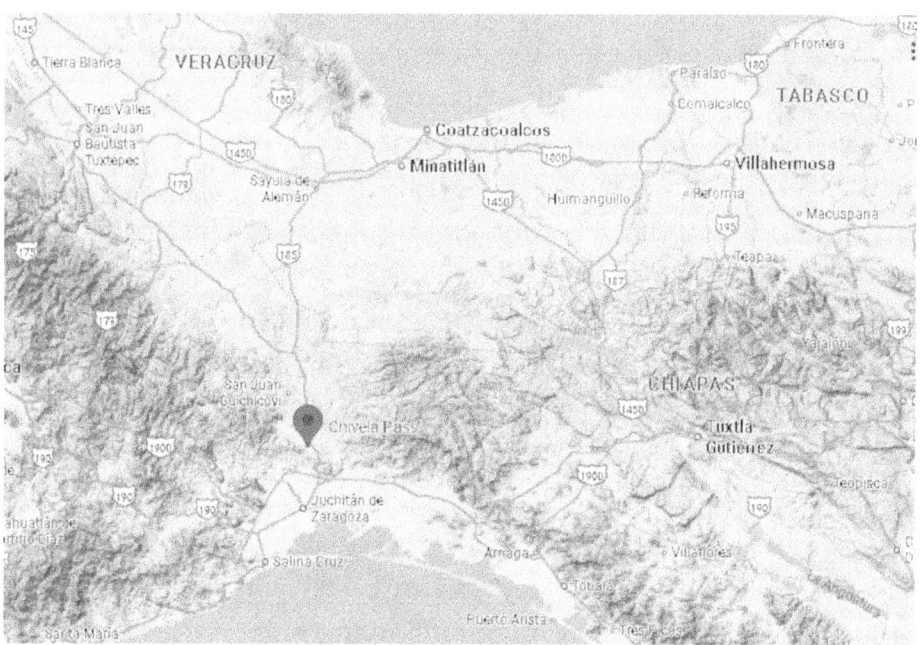

To be honest, people were just lucky anyone picked the Tehuantepec region at all for the region of Book of Mormon lands. As it turns out, the Tehuantepec region is the right place, but not because the lands are in this configuration with seas flanking either side. It is true that the word "narrow" seems to indicate that some sort of landform is narrow in relation to other landforms, which logically will produce some kind of **"hourglass-shape" configuration** in the land. But because it turns out that our candidate is not an isthmus, the "hourglass-shape" is manifested a different way in the real world from the usually-expected way. The reason people that favor Tehuantepec geographies got lucky is because the archaeological **Olmec Heartland** happens to be in the Narrow Neck Region and the Olmecs are our candidate for the Jaredites.

Understanding Book of Mormon Geography

We will give a view of the actual neck from the viewpoint of a map of the **Olmec Trade Routes** going through the pass, and you will be able to visualize the classic, expected "hourglass-shape" that people expect to see from the Narrow Neck of Land. But instead of it being flanked by seas, as I have said, it is flanked by mountains. Here is the bird's eye view of a crude representation of the hourglass shape following the contours of the land, in terms of the area covered by Olmec trade routes through the pass:

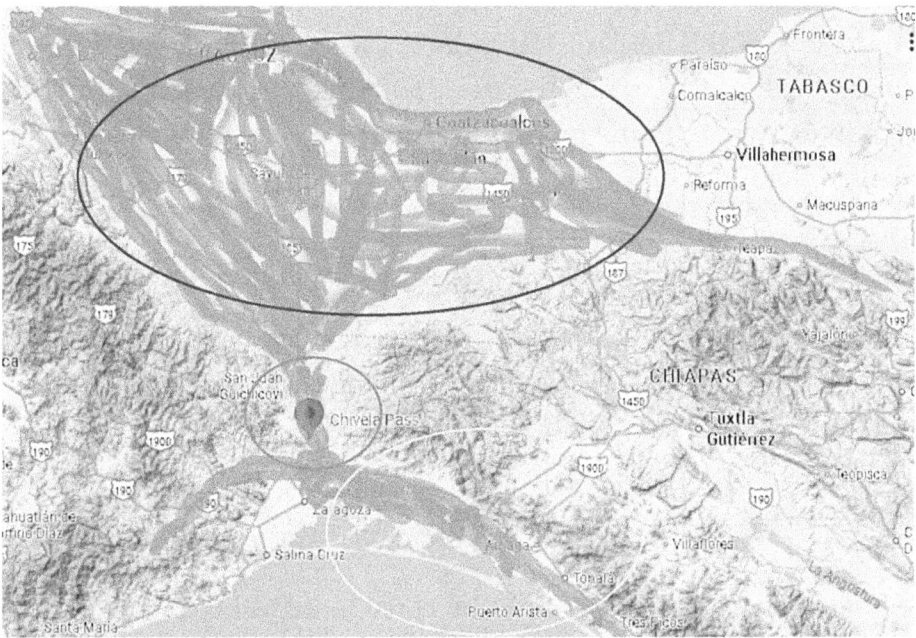

The orange-highlighted areas represent more or less the area covered by the **Olmec Trade Routes** through the area, somewhat following the contours of the land. Using these trade routes as a visual tool is important, because it shows where people were traveling through the area, and how the features of the land fit with what is described in the text of the Book of Mormon. The blue-encircled area represents more or less the Olmec heartland area on the Gulf coast of Tehuantepec. The green-encircled area represents the Chivela pass through the split in the mountain ranges to the east and to the west, with the trade route going through it. The yellow-encircled area shows the part of the Olmec trade routes going along part of Hauck's coastal corridor on the Pacific side. And along this corridor, you can see a series of lagoons and estuaries, which are our candidate for what is referred to in the Book of Mormon text for **where the sea divides the land**, which is a feature mentioned in conjunction with the Narrow Neck of Land/Narrow Pass/Passage.

So, all of this put together shows all of these features in the area, along with the Hourglass shape of the Neck, so that the reader can visualize the features as they are in the real world that are described in the text of the Book of Mormon. As we go in the book, we will focus on all of this in much greater detail in various parts. But this summary here with the maps will help the reader understand what we are talking about as we go.

Methodological Principles: The Role of Evidence

My approach to Book of Mormon geography emphasizes a rigorous methodology, carefully discerning between different types of information and their evidential weight.

Acceptable Archaeological Evidence

For archaeological evidence to be considered valid in this discussion, it must meet stringent criteria:

Academic Acceptability: Evidence must be professionally excavated and documented, not merely based on anecdotal reports from amateur finds. This implies disagreement with individuals who use "questionable artifacts" or "pseudoscientific methods."

Scientific Dating: Artifacts and sites must be scientifically dated (e.g., via carbon dating) to confirm they fall within the Book of Mormon time periods.

Controlled Conditions: Excavations must adhere to professional archaeological standards, ensuring proper context and minimal disturbance.

This means rejecting what I term "**questionable artifacts**" or "**pseudoscientific methods**" often found in non-academic publications. Items like the Los Lunas Stone, the Soper/Savage Collection, or the metal plates from Manti, Utah, regardless of their popular appeal, hurt credibility if they lack proper scientific vetting. Even if a rare few might one day be validated, it is unproductive to build arguments on them now; it is easier to use things that are already accepted. My focus is on already-accepted academic archaeology.

The Place of Historical "Prophetic" Statements

A central point of divergence from the Mesoamericanist paradigm is my view on the role of historical statements from early Church leaders and members. My view also diverges from that of the Heartlanders on this point to a degree because of careful and important refinements in my methodology after the over two decades of time since the time I was a Heartlander, although my approach still has a lot of commonality with theirs. Some scholars assert that only the Book of Mormon text itself can determine geographical locations, dismissing historical statements as unreliable or conflicting. They argue that considering such statements as "evidence" implies an "inspired knowledge" that is not binding on the Church as a whole. While it is true that no *official, binding revelation* on geography has been presented to the Church as a whole, this does not preclude the possibility of key fragments of truth having been revealed to various individuals in the past. Joseph Smith himself believed the area covered by Zion's Camp (Illinois and Ohio) was once inhabited by Nephites.

Joseph Smith himself had the Book of Mormon text in his head during the translation process before it proceeded forth from his mouth, onto the pages written by scribes. It is difficult to believe that he did not have some mental model of where these things took place. This is much different from other prophets that likely did not have such insights. I have had this same belief and assumption since my earliest days in my explorations in Book of Mormon Geography. And this remains a central pillar of Heartlander thought to this day. The difference between my reasoning and regular Heartlander reasoning on this point is because I found information that is not yet well known in Heartlander circles that led to my change of beliefs on the subject of geography after the time of my initial partnership with Wayne May.

My methodology therefore acknowledges that certain historical statements can provide a "key" to understanding the Book of Mormon's geography, in the form of providing real-world anchors. Just as Joseph Smith realized that answers to religious questions had to be sought outside the Bible through direct revelation, so too may geographical keys exist outside the scriptural text itself in these historical records. I employ what has been termed a "dialectical method," where historical statements are used to interpret and harmonize with the Book of Mormon text, rather than dismissing them outright. John Clark called this method "dialectical" because that word indicates an interplay between two sets of data. That idea of interplay between two distinct sets of data is reflected in what John Clark says, that it is "working dialectically between the text and a specific, real-world geography." While some like Clark claim this method leads to "fallacious reasoning" and "compromise of textual and historical details," I contend it's a necessary approach given the ambiguities of the text, and the tendency towards various theorists just interpolating mentally anything they can dream of into the text where those ambiguities lie. It is critical to recognize that while scholarly methods are important, they themselves can sometimes promote fallacious reasoning and compromise of textual and historical details if they are also applied dogmatically to an extreme. Clark's critique of the dialectical method, while valid in some contexts, can become circular, arguing against using real-world sites to test the text because of a lack of certainty about those sites. Yet, Clark and other Mesoamericanists actually break their own rule by mingling archaeological concerns with the data in the text. Furthermore, Clark also acknowledges such a method would be acceptable if certainty already existed on external information to the text in historical accounts. However, since Mesoamericanists apply their approach of avoiding use of any historical/traditional geographic placements amount to a type of extremism. And in some cases, over many years, I have detected a certain smugness and pride in their use of this approach. Therefore, overall, I view this extremism as simply unhelpful. It is a rather self-serving and sanctimonious thing, where Mesoamericanist advocates assert their methodology as superior to all others. This comes not out of revealed knowledge, nor does it come from certainty from anything that suggests this is the case in the text. Ultimately, it emerges only from tribalism and just plain being strongly-opinionated and over-confident because there are so many among their number that have academic credentials. Furthermore, they quote each other in their writings on various points past scholars in their circles have made, ad-nauseam.

While recognizing that historical statements can contain valuable insights, it is important to filter them carefully. Firstly, for reasons I mentioned above, we concentrate only on those most plausibly and clearly from the Prophet Joseph Smith, in order to strain out those from subsequent prophets probably did not have his level of insight on the Book of Mormon text. This is not to denigrate the insights had by other subsequent prophets. But we have to draw the line somewhere because of so much clear contradictions between various statements that could plausibly be called "prophetic." Next, only those statements that harmonize with the Book of Mormon text and our chosen foundational geographical anchors should be considered likely to be inspired; others can be discarded. Some earlier Heartland theories, aiming to place all Book of Mormon geography within the U.S., prioritized seeming "prophetic" historical statements that seemed to support a U.S.-centric Land Southward (e.g., Manti in Missouri, Zarahemla in Iowa). The statements about the Missouri Manti and Zarahemla in Iowa proved to be very problematic, which became evident only in the time periods after the time I was involved in the first *This Land* book with May. However, this narrow focus and insistence on placing things within the borders of the United States is inconsistent with the broader textual evidence and "prophetic statement" evidence now incorporated into our current model, which places the Land Southward in Mesoamerica.

As we have seen, some critics argue that using historical statements for interpretation at all is flawed, leading only to personal speculation. They contend that only the Book of Mormon text alone is sufficient to concretely place its geography on a modern map. But as we have seen, because of the ambiguities, the text *by itself* lacks the explicit "key" for external placement. This inherent ambiguity has indeed led to a multitude of conflicting geographical theories, each filling the gaps with its own preconceived notions. Even the Mesoamericanists rely on archaeology and carbon dating to fill in some of these gaps. Therefore, as the reader can see, I contend that embracing the insights found in certain historical statements offers a path to closer understanding, rather than relying solely on individual textual interpretations. As John Clark observed, "short of receiving pure revelation on the matter, one cannot choose among the geographies based upon what one feels are the relative insights of each, or on the relative completeness of each picture, because each will yield the same number of insights and be approximately of the same caliber." However, this is the truth anyway. Each theorist cannot prove that the set of assumptions that underlie their placements on their maps are any better than anyone else's. This underscores the need for a limited amount of carefully screened external data to serve as a guide. The comparison to interpreting the U.S. Constitution or the Bible, where external writings (like the Federalist Papers) or later texts aiding in interpretation provide crucial "keys" to understanding a text, illustrates this methodological need. Just as scrambled computer data cannot be made meaningful without an external key to the data, Book of Mormon Geography is undecipherable without external keys of some sort in order to get past the clouds of bewilderment created by the ambiguities in the text.

Faith, Bias, and Interpretation

It is important to acknowledge that all Book of Mormon geography theories, including my own, are ultimately "faith-based" to some degree, being put forth by religious individuals making religious interpretations of archaeological and scientific evidence. An apologist, by definition, starts with a belief and seeks to defend it, employing methodologies that differ from pure scientific inquiry. However, this does not invalidate the arguments, except in the view of pure secular rationalists. But because the impetus in making an argument for a real-world placement for the Book of Mormon is religious in the first place, we aren't really looking for secular rationalists to be the primary audience for this kind of argumentation in the first place, and it isn't really my concern to seek after their approval anyway. Since a broader discussion with those types of people actually revolves around the root issue of the historicity of the Book of Mormon anyway, and geography is just a peripheral issue anyway, then they shouldn't concern themselves to much with an issue that builds upon a bias based on personal testimony that the Book of Mormon is true in the first place. Rational arguments serve to "maintain a climate in which belief may flourish."

My approach recognizes the presence of **confirmation bias** in all geographical interpretations. As we discussed earlier, Restorationist scholars of any stripe that claim to avoid the "dialectical method" (interpreting text through external data) and rely only on the text, they often unconsciously replace historical statements with their *own* preconceived notions, reading their preferred geography into the ambiguous textual descriptions anyway. This leads to varied interpretations. What one geographer hobbyist thinks that one thing is plain and absolutely clear, another will think it is ambiguous and unclear. This just proves that ambiguities are all over the place. This is why the method of relying solely on internal textual analysis without external anchors has not yielded definitive results for Mesoamericanists over decades. They have simply just proven that they have reinforced over and over again their own cognitive biases. Some have asserted that the Mesoamerican thesis has the bulk of textual and scientific evidence on its side, but I argue that this only represents the bulk of Mesoamericanist scholarly opinion about what the text means, rather than unequivocal, unambiguous and definitive results.

My personal journey with this topic involved wrestling for decades with Mesoamericanist arguments, often experiencing "cognitive dissonance" when their claims contradicted my strong conviction about the New York Cumorah. I found that rationalizing away what I felt to be true consistently led to a "troubled feeling" that only dissipated when I re-embraced my conviction. This internal struggle shaped my approach, leading me to trust my own perceptions and experience, even if they sometimes diverged from academic consensus, while still acknowledging the need for rational defense. Yet, whether it is personal testimony reflecting actual truth, or just merely strong personal conviction, it is still at the root of what all geography hobbyists build upon. It is a cognitive bias that colors everything that each hobbyist does. Mesoamericanists are as guilty of it as anyone, and as unable to not be led by it as anyone else. Yet, some, like John Clark warn against this type of thing as a guide in scholarly study or argumentation, as if somehow the Mesoamericanists have avoided it. Speaking of this issue in a review of some material produced by Book of Mormon Geographer Delbert Curtis, he writes:

> I do not doubt that Curtis sincerely believes his claims, but his beliefs are not binding on anyone else. It is poor practice to

accept lay testimony as fact, and I will not do so here. The entire history of the Church, and my personal experience with numerous peoples' personal witnesses concerning the location of the Nephite repository of gold plates, suggests that we should treat such diverse and contradictory testimonies with extreme caution. Here, I do not consider the evidence of personal testimony as relevant to scholarly argument.[1]

Of course, the shared baseline or consensus of ALL believing Book of Mormon scholars, whatever their faction, is personal testimony in the truth of the historicity of the Book of Mormon. Testimony is where an argument in favor of anything positive about the Book of Mormon begins for believers. If it were truly poor practice to accept lay testimony or just plain conviction as fact in any scholarly argument in all cases, then the whole Church should just give up on the Book of Mormon, disband the Church, and we should all go join protestant or catholic religions, as absurd as that would be.

Therefore, my methodology integrates multiple lines of evidence:

Book of Mormon Text: The primary source, interpreted carefully for its internal geographical relationships and directional cues.

Academically Accepted Archaeology: Scientifically dated and professionally excavated evidence from relevant cultural periods (e.g., Olmec, Hopewell-Adena).

Historical Statements and Traditions: Accounts from early Church leaders and members, viewed as potential "fragments of truth" or "keys" that can unlock the meaning of the text's geography.

Indigenous Legends: Oral traditions and historical records from Native American tribes, which may contain echoes of ancient events relevant to the Book of Mormon.

This comprehensive, multi-faceted approach, while acknowledging its faith-based foundation, seeks to build a coherent and ratio-

---

1. Review of Books on the Book of Mormon, Vol. 6 (1994), No. 2, p. 80

nally defensible model for the geography of the Book of Mormon. This requires recognizing that the intellectual landscape of Book of Mormon geography is often prone to overconfidence, dogmatism and strongly-opinionated words from various theorists, despite the inherent complexity and lack of definitive evidence for many aspects. As John W. Welch advises, scholars should exhibit humility and avoid dismissing alternative plausible explanations simply because they do not fit a favored paradigm. The "answer no can never be quite as wrong as the answer yes" in unproven scholarly assertions, as Hugh Nibley reminds us. And rejecting an idea purely on "personal incredulity" or "scholarly consensus" (argumentum ad populum) rather than substantive counter-evidence is a logical fallacy that hinders progress. Thus, this book aims to provide a rational framework for a belief in the New York Cumorah that withstands scrutiny, by employing the combined strength of datasets from both the United States heartland, as well as that of Mesoamerica. Acknowledging that even if one remains in the minority is paramount. When one fights for one's personal conviction is an important contribution to a larger intellectual environment that helps support the truth of the Book of Mormon in general, and may require personal sacrifice. But in the end, a rationally defended belief that is published in this marketplace of ideas may end up serving and helping others in various ways to lend strength to their core convictions, and thereby, help them in the end to choose to keep their covenants with God and not fall away. I have always wanted to solve the puzzle of Book of Mormon Geography, not just out of curiosity, but to support faith in the actual core truth claims of the gospel.

Chapter 2

Lehi's Landing: Traditions and Possibilities

The Book of Mormon begins with the departure of Lehi's family from Jerusalem around 600 BC and their subsequent journey across the ocean to the promised land. While the text is rich in spiritual and historical detail, it is notably sparse on precise geographical locations for their landing and subsequent early migrations. This ambiguity has led to various theories and traditions over the centuries regarding Lehi's initial point of arrival in the Americas. Understanding these traditions, and evaluating them against geographical and scientific insights, is crucial for a comprehensive approach to Book of Mormon geography.

One prominent traditional theory, a central facet of the Hemispherical theory, regarding Lehi's landing place points to **Chile, in South America**. This idea gained traction in early Latter-day Saint thought and was even expressed in semi-official capacities. A significant account supporting this tradition comes from **Frederick G. Williams**, one of the early leaders in the Church. His account suggests a landing in Chile, specifically "thirty degrees south latitude." This tradition likely found favor because Chile, being a significant landmass on the Pacific coast of South America, could logically be reached by a sea voyage from the Old World, especially one that might have rounded Africa and then navigated across the Pacific.

Further reinforcing the Chile tradition are statements from early

Mormon missionaries, which reflect the expansive geographical understanding prevalent in the early Church. For instance, an early report on LDS missionary activity in Ohio, dated February 16, 1831, broadly stated: "The whole Continent from Chili to Canada, has been the scene of wars, bloodshed, and carnage for a period of 1000 years, according to the account of Mormon, and during this time, and at the time of their final destruction, the Messiah, or as the Book calls him, the Christ, visited the inhabitants and established his church, etc."[1] This quote, though not explicitly detailing Lehi's landing, demonstrates an early belief in a hemispheric scope for Book of Mormon events that clearly included Chile.

The geographical appeal of Chile for a trans-Pacific voyage is also bolstered by **oceanic currents**. The **Humboldt Current**, also known as the Peru Current, plays a significant role in Pacific oceanography. This cold, low-salinity ocean current flows north along the western coast of South America from the southern tip of Chile to northern Peru. While Lehi's journey would have likely approached from the west across the Pacific, the presence of such a strong northward-flowing current along the coast of Chile and Peru could have been a factor considered by early theorists. A vessel catching favorable winds and currents could be drawn towards this region. Therefore, the Humboldt Current, though not a direct route *to* Chile from the West, explains the prevailing oceanic conditions along that coastline and might have been perceived as facilitating navigation for ancient mariners once they reached the South American landmass.

However, another historical tradition and an intriguing modern interpretation point towards the **Isthmus of Darien**, located in present-day Panama and Colombia. This area, known for its dense

---

1. https://tinyurl.com/38dh9n28 ; A. S., "The Golden Bible, or, Campbellism Improved," *Observer and Telegraph* (Hudson, Ohio) (November 18, 1830); See Larry E. Morris, *A Documentary History of the Book of Mormon* (New York, NY: Oxford University Press, 2019), pp. 384–386

jungles and challenging terrain, represents a narrow strip connecting North and South America.

Significantly, there are indications from early Church history that **Joseph Smith or John Taylor** alluded to a possibility of Lehi's landing near, or a little southward of, Darien (the isthmus of Panama).[2] The *Times and Seasons* (a Church publication of which Joseph Smith was editor) published articles that identified this area and other areas in Central America as potential Book of Mormon lands. While a direct quote unmistakably and unambiguously from Joseph Smith explicitly stating "Lehi landed in Darien" might be elusive, this holds great weight showing that this idea was at least in the minds of people that were in the inner circle of the Prophet, and the idea may have been in his, seeing that he was editor of this, as we stated. Prominent Heartlanders and others have argued against this, but at least this shows that the idea existed in the minds of prominent early saints.

Orson Pratt and others believed in the Williams so-called "Revelation" alleging that Lehi landed in Chile. Orson Pratt wrote, "They . . . landed on the western coast of South America. . . . "[3] "[The] Nephite nation about this time commenced the art of shipbuilding. They built many ships, launching them forth into the western ocean. The place of the building of these ships was near the Isthmus of Darien."[4] This idea that Hagoth was launching ships from Panama of course was based on the earlier speculation about Lehi's landing in Chile, because Panama was the "obvious" Narrow Neck of Land to them. The Magdalena River was an "obvious" candidate for the River Sidon for being the main major river that ran northward, just south of Panama. As it turns out though, the place of Lehi's landing, has not been revealed:

---

2. There are many publications on this issue. See for example Jeff Lindsay's: https://tinyurl.com/mvprurfk *Book of Mormon Nuggets: Nugget #11: What Could Joseph Smith Have Known about Mesoamerica?*
3. *Journal of Discourses* 13:129-31
4. *Journal of Discourses* 14:325-30, 333

> The Church has issued no information concerning the route followed by Lehi.... Until this is done, teachers of the Gospel Doctrine department should refrain from expressing definite opinions.... The present associate editor [George D. Pyper] of The Instructor was one day in the office of the late President Joseph F. Smith [who died in 1918] when some brethren were asking him to approve a map showing the exact landing place of Lehi and his company [i.e. in South America]. President Smith declined to officially approve the map, saying that the Lord had not yet revealed it, and that if it were officially approved and afterwards found to be in error, it would affect the faith of the people.—Asst. Editor.[5]

Interestingly, though, the older edition of the book *Articles of Faith* by James Talmage says the following about Lehi's party:

> Their voyage carried them eastward across the Indian Ocean, then over the South Pacific Ocean to the western coast of South America, whereon they landed.... They spread northward, occupying the northern part of South America.

It seems to say this rather matter-of-factly. Since that time, the General Authorities have purposefully changed that in the book to reflect what appears to be the current position of the church on the subject, at least at that time, thus leaving the statement in the original version clearly superseded. No doubt, they did this because they could not leave such a quasi-official book as that in such a state, when this geographical setting is not revealed. It is very interesting what the newer revision (at least the 1981 edition) has in its place (and the wording has a lot less certainty than the original,

---

5. Sorenson, John, *The Geography of Book of Mormon Events: A Source Book*, Provo, UT: FARMS, 1992, p. 387

because of the use of the terms "believed," "traditionally believed" and so forth:

> It is believed that their voyage must have carried them eastward across the Indian Ocean, then over the Pacific to the western coast of America, whereon they landed about 590 B.C. The landing place is not described in the book itself with such detail as to warrant definite conclusions. . . . It is traditionally believed that they spread northward, occupying a considerable area in Central America.[6]

It is interesting that this would point to the information coming from the *Times and Seasons* articles, suggesting that the area of Zarahemla was likely to be in Mesoamerica, and that Lehi landed just south of Darien!

It is not very clear exactly when the change in wording happened, because I have not comprehensively tried to look at every edition of the text of *Articles of Faith* (because comprehensiveness on this one minor point is not necessary). Nevertheless, this change most likely was made long *after* the death of James Talmage, the author of the book. And it only would have happened because of the change in popularity of the idea that Mesoamerica was the seat of the Nephite civilization, being at the very least, its Land Southward, and the awareness of this among the General Authorities. This change in wording highlights the fact that it is nothing more than a traditional belief that Lehi landed in Chile, and such a definite conclusion is not warranted from the text. But it also highlights that it is *also* a tradition that that Lehi landed just south of Darien, so there are two conflicting traditions here. Nevertheless, the tradition that was chosen and favored for the newer version of the text, as of 1981, **was that of the Mesoamerican traditional theory from the *Times***

---

6. Talmage, James, *Articles of Faith*, Salt Lake City, Utah: Deseret Book Co., 1981, pp. 258-259, emphasis added.

and Seasons **articles.** It highlighted the tentative nature of it, that it was a mere belief. It also highlights the fact that the Mesoamerican theory for the Land Southward is also a mere belief and tradition. Nevertheless, a change in such a quasi-official book like this is not made lightly, and had to have been made authoritatively. Someone in the leading quorums of the Church authorized the changing of the text! Such a thing could have happened no other way. It is somewhat of a window into the thinking of certain brethren at the time, showing what some apparently favored at that time.

Significantly, an account exists that was mentioned by Albert James Pickett in his publication about the history of Alabama, describing a tradition among the Native Americans of a crossing of the Pacific and a landing near Darien and subsequent migration northward over time, ultimately into the present area of the United States, of the Creek Indians (the Muscogee tribe):

> In 1822, Big Warrior, who then ruled the Creek confederacy, confirmed this tradition, even going further back than Milfort, taking the Muscogees from Asia, bringing them over the Pacific, landing them near the Isthmus of Darien, and conducting them from thence to this country. "My ancestors were a mighty people. After they reached the waters of the Alabama and took possession of all this country, they went further—conquered the tribes along the Chattahoochie, and upon all the rivers from thence to the Savannah—yes, and even whipped the Indians then living in the territory of South Carolina, and wrestled much of their country from them." The Big Warrior concluded this sentence with great exultation.[7]

As we see, this traditionary evidence directly coincides with the

---

7. https://tinyurl.com/4jzr2ky5 ; Albert James Pickett, *History of Alabama and Incidentally of Georgia and Mississippi, From the Earliest Period*, Chapter 3, emphasis added

geographical framework that was suggested in the *Times and Seasons* articles from early in this dispensation that we have been discussing that feature the ancient American ruins of Mesoamerica and attributes them Book of Mormon peoples.

My current model, the Dual-Heartland Theory, primarily places the **Land Southward (Zarahemla and Nephi) in Mesoamerica** and the **Land Northward (Desolation, Hopewell regions) stretching into the Midwestern United States and New York**. Given this framework, the question of Lehi's initial landing becomes even more critical. If Lehi landed in Chile, a significant subsequent journey would have been required to bring his people to the core Book of Mormon lands in Mesoamerica. But that is not implausible or unlikely.

This leads to an interesting possibility: **Lehi's party may have first arrived along the coast of Chile as the first place they got to. And perhaps they may have even landed there as a first stop. There may have even been a bit of initial exploration there.** If so, after this brief period, Nephi, under the direct guidance of God communicated through the Liahona, perhaps received instruction to continue their journey northward. If so, this divinely directed continuation led them along the coast in the boat, where they were then instructed to land a little south of the Isthmus of Darien. This scenario allows for the historical traditions of a Chile landing while also providing a divinely guided transition to the Central American region, which subsequently became the primary stage for the early Book of Mormon narrative. The Humboldt Current, with its initial eastward flow, and then a northward flow along the coast of South America, could have naturally facilitated this divinely orchestrated continuation of their voyage along the coastline towards their ultimate prophesied destination south of Darien.

It is worth noting that some earlier theories, primarily focused on a U.S.-centric geography, proposed that Lehi's land of first inheritance could have been either in Louisiana, Florida, or in the

St. Lawrence Seaway. All of these ideas, however, face significant geographical and textual challenges when considering the internal geography of the Book of Mormon, and have no traditionary support like the account of landing near Darien. Cumorah is consistently placed in the Land Northward, while Lehi's landing was in the Land Southward, far from the Hill Cumorah. Therefore, the textual and geographical constraints, as well as the traditional account quoted above, make a landing somewhere in the Great Lakes region unlikely for Lehi's initial arrival.

Chapter 3

Joseph Smith's Geographical Insights and the Hopewell Connection

Joseph Smith, the first prophet of the Latter-day Saint dispensation, consistently evinced a profound understanding of ancient American inhabitants, particularly the Nephites, well beyond what could be attributed to mere speculation. His personal statements and observations provide crucial insights into the geographical setting he envisioned for the Book of Mormon, especially regarding the North American continent, and specifically the area of it now comprising the United States.

His mother, Lucy Mack Smith, recalled how Joseph would vividly describe things regarding the ancient peoples:

> During our evening conversations, Joseph would occasionally give us some of the most amusing recitals that could be imagined. He would describe the ancient inhabitants of this continent, their dress, mode of traveling, and the animals upon which they rode; their cities, their buildings, with every particular; their mode of warfare; and also their religious worship. This he would do with as much ease, seemingly, as if he had spent his whole life among them.[1]

---

1. Smith, Lucy Mack, 1954, 1901, *History of the Prophet Joseph Smith*, Salt Lake City, UT: Bookcraft, p. 83

To emphasize, again, it was done "with as much ease, seemingly, as if he had spent his whole life among them." This intimate knowledge, extending to the specifics of their "cities and buildings," strongly implies his likely awareness of their geographical placement.

Beyond these general recollections, Joseph Smith made direct statements linking Book of Mormon peoples to the Midwestern United States. During Zion's Camp in 1834, as his company traveled through Ohio and Illinois, he wrote to his wife Emma:

> The whole of our journey, in the midst of so large a company of social honest and sincere men, wandering over the plains of the Nephites, recounting occasionally the history of the Book of Mormon, roving over the mounds of that once beloved people of the Lord, picking up their skulls and their bones, as a proof of its divine authenticity, and gazing upon a country the fertility, the splendour and the goodness so indescribable, all serves to pass away time unnoticed.[2]

This is a clear, first-hand account indicating Joseph's opinion that the Nephites inhabited the Midwestern United States, specifically the area from eastern Ohio to Illinois, and that they built the mounds in that region.

This connection between the mounds of the Midwest and the Book of Mormon was not confined to Joseph's private correspondence. Early missionaries openly preached this correlation. For example, David Marks, while visiting the Ohio mounds, was told that "the 'Book of Mormon' gave a history of them, and of their authors." He subsequently became anxious to get a copy.[3] A re-

---

2. Letter to Emma Smith, June 4, 1834, in Jessee, Dean C., *The Personal Writings of Joseph Smith*, Salt Lake City, Utah: Deseret Book Co., 1984, p. 324, spelling corrected

3. Vogel, Dan, Indian Origins and the Book of Mormon, Salt Lake City, Utah: Signature Books, 1986, p. 32

port from Boston in 1834 noted that the LDS people "Suppose the mounds throughout the western states, which have heretofore excited so much curiosity, are the remains of the cities of the Nephites and Lamanites."[4] Similarly, Edward Strut Abdey wrote in 1835 that "the mounds of earth . . . are referred to, by the preachers of the Mormon faith, as proofs of these theocratic tribes."[5] An 1841 pamphlet by LDS elder Charles Thompson asserted that the similarities between the mounds and the descriptions in the Book of Mormon "were sufficient to show to the public that the people whose history is contained in the Book of Mormon, are the authors of these works."[6] These observations confirm that the connection between Book of Mormon peoples and the mound-building cultures of North America was a known and taught aspect of early Latter-day Saint belief.

Joseph Smith himself gave other specific identifications that further solidify this connection:

Adam-Ondi-Ahman: In 1838, Joseph named a location in Missouri "Tower Hill," stating it was "in consequence of the remains of an old Nephite altar or tower that stood there."[7] George W. Robinson, a scribe accompanying Joseph, corroborated this, recording: "We next kept [traveling] up the river mostly in the timber for ten miles, until we came to Colonel Lyman Wright's who lives at the foot of Tower Hill. A name appropriated by President Smith in consequence of the remains of an old Nephitish Altar and Tower where we camped for the Sabbath."[8] The observation of scattered large stones on Tower Hill confirms the remains of a significant ancient structure, likely Hopewellian, which Joseph identified as Nephite. Robinson also recorded: "President Smith and myself . . . returned

---

4. Ibid., p. 32
5. Ibid. pp. 32-33
6. Ibid. pp. 32-33
7. History of the Church 3:34-35
8. Scott H. Faulring ed., An American Prophet's Record: The Diaries and Journals of Joseph Smith, SLC: Signature Books, 1989, p. 184.

to the camp in Robinson's Grove . We next scouted west in order to obtain some game to supply our necessities but found or killed none. We [found] some ancient antiquities about one mile west of the camp, which consisted of stone mounds, apparently laid up in square piles, though somewhat decayed and obliterated by the almost continual rains. Undoubtedly these were made to seclude some valuable treasures deposited by the aborigines of this land."[9]

The Zelph Incident: As detailed in chapter 5, the discovery of Zelph's skeleton in an Illinois mound in 1834 led to Joseph Smith receiving a vision identifying him as a "white Lamanite" and warrior "under the great prophet Onandagus, who was known from the Hill Cumorah, or eastern sea to the Rocky mountains."[10] *This incident directly links Book of Mormon peoples to the Hopewell mound-building cultures and geographically anchors them in the Midwestern United States.*

These statements, whether direct pronouncements by Joseph Smith or corroborating accounts from his associates, clearly demonstrate that in his opinion, Book of Mormon events and peoples extended into the heartland of the United States. While Heartlander theorists currently maintain (and I used to maintain) that they were indicative that Joseph Smith believed in the heartland model, when relying on these statements alone, the evidence would seem to be ambiguous about which model that includes these areas ought to be favored. Because while establishing that the Nephites were in the area, nothing else can really be concluded just based on this alone. Even the Hemispherical model maintained the conviction that the Nephites inhabited the region of the Midwestern United States. This is why it is not uniquely supportive *only* of the idea

---

9. Faulring, Scott H. ed., 1989, An American Prophet's Record: The Diaries and Journals of Joseph Smith, Salt Lake City, Utah: Signature Books, p. 185, spelling corrected

10. History of the Church, Vol 2:5:79-80

that Joseph Smith may have supported a model like the Heartland Model.

In this current model proposed in this book, the Dual-Heartland Model also includes this area, maintaining along with the other models that include this area, that this ancient, urban area of the United States is indeed one of the "heartlands" of the Nephite Nation, notwithstanding I also maintain that the urban centers of Mesoamerica comprised the southern "heartland."

Notwithstanding the existence of the Hemispherical model in the 19th century, BYU scholar Andrew Hedges provided an analysis of Joseph's probable "limited hemispheric" or "northern hemispheric" view about which he wrote in his writings. According to Hedges, the documentary evidence of the time of Joseph Smith, instead of supporting a Hemispherical model very well, actually showed that the Book of Mormon peoples inhabited parts of the eastern United States, with final battles in New York, and civilizational centers in Mesoamerica. This broad observation aligns with the essentials of my current model. Hedges wrote:

> The purpose of this paper is to identify the general view or model of Book of Mormon geography that emerges from a careful review of these eleven documents. I do this not in an effort to identify Joseph's particular views on the topic (which I believe is impossible to determine; see below), but to better understand the picture of Book of Mormon geography that was generated as Joseph and his close associates worked to fulfill his responsibilities as newspaper editor, correspondent, historian, and defender of the faith. A review of these documents also helps us understand what Church members were reading on the topic in Church sponsored publications, where the most explicit of them were published. I argue that while these particular documents paint a picture of an extensive geography for Book of Mormon events, none of them, separately or in

the aggregate, necessarily suggest the fully hemispheric view of the book's geography that Sorenson suggests most Church members held. Nor can any of them, or any combination of them, be interpreted to support the idea that the author(s), and perhaps Joseph Smith himself, necessarily envisioned a limited Mesoamerican geography for the book's events, as John Clark has recently suggested. At least one of them, in fact, clearly indicates on its own, without being viewed in the context of the others, a belief in a geography extending from Central America up into the Ohio River Valley. This same document and similar documents, written less than two years before Joseph's death, may also require us to qualify Terryl Givens' suggestion that the "efforts of Joseph and his brethren to identify Book of Mormon lands would increasingly focus southward" over time. Rather than a hemispheric or limited geography, or some sort of development from one to the other, the view of Book of Mormon geography contained in this particular subset of documents is one that has Book of Mormon peoples, during at least part of their histories, inhabiting parts (although not necessarily all) of the eastern United States; the final battles of the Nephites and Jaredites taking place in upstate New York; and the centers for both the Nephite and Jaredite civilizations being located somewhere in Central America. South America is largely, if not completely, out of the picture, while sites thousands of miles apart in Central and North America are very much in. . . .

The documents reviewed in this paper suggest an understanding of Book of Mormon geography lying somewhere between a fully hemispheric model, on the one hand, and a limited model on the other. According to this view, which we might dub a "limited hemispheric" or "northern hemispheric" view, Book of Mormon peoples, during at least part of their histories, inhabited parts (although not necessarily all) of the eastern United States; the final battles of the Nephites and Jaredites took place in upstate New York; and the centers for both the Nephite and Jaredite civilizations were located somewhere in Central America.[11]

While many BYU scholars predominantly favor a Mesoamerican perspective, my analysis offers a contrasting view regarding the true extent and nature of ancient Book of Mormon civilizations. Hedges's historical analysis, though focused on the correct geographical area for the Nephite nation, differs from the archaeological evidence presented here. His work, being primarily historical, did not delve into the archaeological record of the eastern United States as this book does.

Hedges suggested that the "centers for both the Nephite and Jaredite civilizations were [primarily] located somewhere in Central America," with Nephites merely "inhabiting" parts of the United States. However, the archaeological record strongly indicates a high degree of urbanization and a substantial, albeit more dispersed, population within the eastern United States, particularly in areas of Ohio and Illinois. This contrasts sharply with the idea of the region being sparsely populated or merely a "hinterland."

Some critics may attempt to dismiss this conclusion as confirmation bias stemming from my past "Heartlander" views. This is inaccurate. My findings are the only truly objective conclusion reached after a careful and thorough review of the archaeological data from the region.

Unfortunately, scholars advocating a Mesoamericanist viewpoint have consistently overlooked the profound implications of this archaeological evidence. I have meticulously studied this archaeological picture since my time as a Heartlander, and the Mesoamericanists' analyses have consistently missed the mark, including their

---

11. Hedges, Andrew, "Book of Mormon Geography in the World of Joseph Smith" https://tinyurl.com/4ct3symh

flawed interpretations of the archaeology surrounding Cumorah in New York. This is as a result that they are specialists in other areas. Almost none are specialists in North American archaeology. Even those who concede that the eastern United States was a mere "hinterland" of Book of Mormon society inaccurately suggest it was a sparsely populated area lacking significant population centers. The true scale of the Book of Mormon heartland in the North, supported by archaeological findings, indicates the opposite. Just because the population was greater in Mesoamerica doesn't take away from the grandeur of the ancient society in Ohio and Illinois, with Cumorah in New York in its backyard, which was PART of the same civilization of the South, being its satellite.

The Hopewell Connection: A Thriving Urban Civilization in the Land Northward

Joseph Smith's geographical insights for the Land Northward find compelling archaeological correlation with the **Hopewell tradition** (or "culture") and the closely related **Adena culture**. These ancient Native American peoples flourished in the Eastern United States, primarily from approximately 200 BC to 400 AD,[12] a period that largely overlaps with the Nephite civilization's later centuries and ultimate destruction. Some archaeologists even date the Adena's origins as early as 1500 BC.

The Hopewell tradition was not a single, monolithic culture but a vast network of related populations connected by the sophisticated **Hopewell Exchange System** (also known as the Hopewell Interaction Sphere). This system facilitated trade routes stretching from the Atlantic coast to the Rocky Mountains, and from the Great Lakes to Mexico, diffusing Hopewellian traits over vast areas of the continent.[13] This extensive network demonstrates the plausible scope of

---

12. Hopewell Culture, https://tinyurl.com/3n65sek2
13. Ibid.

interaction and movement that could have connected a Mesoamerican heartland with a northern domain in the U.S. While the regional traditions within this sphere might have appeared distinct, significant ties and shared cultural elements were maintained across this expansive network.

The Hopewellians exhibited numerous characteristics that make them strong candidates for the Nephite and related cultures in the Land Northward:

Advanced Society: They were highly organized, directed by able leaders, with a stratified social structure and specialized guilds of craftsmen and traders—some even calling them "the first craft unions of America." Their activities were well-organized, demonstrating ingenuity in social cooperation for massive public works.

Monumental Earthworks: While Mesoamerican civilizations built extensively in stone, Hopewellians excelled in constructing massive earthen mounds and geometric enclosures. These "impressive" earthworks, often ceremonial centers, are the primary types of ruins left in Eastern North America. Some, like Fort Ancient, appear defensive, while others exhibit astounding symmetry and astronomical alignments, functioning as ceremonial structures or observatories. This contrasts with a prior assumption that North America lacked grand structures, highlighting the diversity of ancient building traditions. Evidence of structures built primarily from wood, which would naturally decay over time, explains what some have believed is mere "paucity of evidence" in some areas. In reality, extensive structures were built on and around the mounds.

Metallurgy: The Hopewellians were "the finest metalworkers of all prehistoric Indians," capable of beating and annealing copper, iron (from meteoric origin), silver, and gold into a variety of tools, ornaments, and even breastplates.[14] Their copper work often involved heating above 600°C for homogenization, with some artifacts showing evidence of heating above 1000°C, indicating ad-

---

14. https://tinyurl.com/yfb9yk4x ; Brittanica, Hopewell Culture

vanced metallurgical techniques. The discovery of cast copper artifacts in the Old Copper Complex of Minnesota further surprised archaeologists, providing "firm evidence of casting"[15] These capabilities align with Book of Mormon descriptions of metallurgy among its peoples, who possessed "swords of steel" (Ether 7:9; Jarom 1:8). Many other descriptions in the Book of Mormon. It's true that we await more evidence for clear instances of steel in the archaeological record that are alloys of *carbonized iron (i.e. the normal kind of "steel" we are used to thinking of)*. This hasn't been found in Mesoamerica either yet. Nevertheless, hardened copper objects are an important evidence

Agriculture: They were competent farmers, cultivating plants like sunflowers, goosefoot, squash, and later corn and beans. Critically, archaeological specimens of domesticated **barley** (Hordeum pusillum) have been discovered in sites across North America, including Illinois and Oklahoma, dating back to the Book of Mormon time period.[16] This directly addresses a common criticism leveled against Book of Mormon historicity due to the mention of "barley" in the text. Corn pollen dated to 2500 BC in Illinois also indicates early cultivation.

Literacy and Record Keeping: While their writing systems differed from Mesoamerican glyphs, Hopewellian peoples, and their descendants like the Iroquois and Algonquians, had sophisticated ways of recording history. The Ojibway had circular copper plates with hieroglyphics marking generations,[17] and the Iroquois used wampum belts as mnemonic devices to tell complex historical narratives,[18] demonstrating a distinct form of literacy.

William Warren an Ojibway historian wrote about a copper plate

---

15. https://tinyurl.com/8euxv44c ; Old Copper Complex
16. https://tinyurl.com/4smurrnu ; FAIR Blog, Another Look at Barley in the Book of Mormon
17. https://tinyurl.com/awajjxn7 ; Kevin E. Smith, *Mississippian Feline Copper Plates from the Southern Appalachian Region*
18. https://tinyurl.com/4wh8r8n5 ; Wampum

with writing on it: "[T]his family hold in their possession a circular plate of virgin copper, on which is rudely marked indentations and hieroglyphics denoting the number of generations of the family who have passed away since they first pitched their lodges at Shaug-a-waum-ik-ong and took possession of the adjacent country, including the Island of La Pointe or Mo-ning-uwn-a-kaun-ing. When I witnessed this curious family register in 1842, it was exhibited by Tud-waug-aun-ay to my father. The old chief kept it carefully buried in the ground, and seldom displayed it. On this occasion he only brought it to view as the entreaty of my mother, whose maternal uncle he was. . . . I am the only one still living who witnessed, on that occasion, this sacred relic of former days. On this plate of copper was marked eight deep indentations, denoting the number of his ancestors who had passed away since they first lighted their fire at Shaug-a-waum-ik-ong. They had all lived to a good old age. By the rude figure of a man with a hat on its head, placed opposite one of these indentations, was denoted the period when the white race [i.e. the Europeans] first made his appearance among them. This mark occurred in the third generation, leaving five generations which had passed away since that important era in their history. Tug-wang-aun-ay was about sixty years of age at the time he showed this plate of copper, which he said had descended to him direct through a long line of ancestors. He died two years since, and his death has added the ninth indentation thereon; making at this period, nine generations since to Ojibways first resided at La Pointe, and six generations since their first intercourse with the whites."[19]

This is very significant that they were writing on metal. Reports among various tribes of symbolic writing on birch bark are widespread, especially among the Ojibway:

"Birchbark scrolls used in the Ojibwa religion (Midewiwin) were

---

19. https://tinyurl.com/ysjk8hve; Warren, William W., *History of the Ojibway People*, St. Paul, Minnesota: Minnesota Historical Society, 1885, pp. 89-90

sewn together in similar fashion, and a form of pictographic writing was inscribed with a pointed instrument. . . . These inscriptions were highly symbolic and could be read only by trained practitioners."[20] Similarly, the Rock Art spread throughout all regions of North America shows the widespread use of symbolic communication, similar to the mnemonic communication in the wampum and on this kind of plate used by the Ojibway. With these types of symbolic communication focusing on using symbols as key memory cues and the use of songs to preserve traditional accounts that go along with the symbols to preserve the whole content shows the lack of necessity for a "running text" kind of written communication in all instances. While we await evidence of "running text" type of writing in the ancient United States, and instances of the Nephite type of writing shown in the Anthon transcript, all of this challenges ethnocentric views that only specific types of written languages constitute "literacy." The Nephites themselves noted that most of their writing on perishable materials (such as on birchbark scrolls) would "perish and vanish away," which explains the scarcity of such records in the archaeological record.

Mesoamerican Influence: Crucially, there is strong evidence of Mesoamerican influence in Hopewell territory. As early as the archaic period, Olmec incursions led to the establishment of sites like Poverty Point in Louisiana (dated from 1115 BC), which shows Olmec influences in its monumental earthworks and imported materials from a vast radius. Scholars like John Sorenson have documented a "long sequence of cultural transmissions and migrations moved northward from southern Mexico," with evidence for cultural and linguistic transmissions from Mexico in the Eastern U.S.. For example, Hopewell corn was a Guatemalan type.[21] The

---

20. https://tinyurl.com/ye3fpbck ; Earl Nyholm, *The Use of Birch bark by the Ojibwa Indians*

21. https://tinyurl.com/52sczy8e ; https://tinyurl.com/5cdjdcmk ; Chapter 63, entitled "Mesoamericans in Pre-Columbian North America," in the book *Re-exploring the Book of Mormon*

presence of a "Great Hopewell Road" in Ohio, a "Mesoamerican sacbe" or "White Road," archaeo-astronomically linking sites, further demonstrates profound Mesoamerican ties. This "White Road" concept in Mesoamerican thought is deeply tied to ethnoastronomical beliefs about the Milky Way as a "cosmological road," parallels which are also found among North American tribes like the Wyandots and Lakota. This indicates a shared ideological framework, not just casual trade.

This quote demonstrates the level of sophistication in the astronomical knowledge that came forth from Mesoamerica to the Great Lakes Region:

> Hively and Horn (1982) added a significant dimension to our appreciation of the Hopewellian achievement when they determined that the major rising and setting points of the moon, encompassing an 18.6 year cycle, are incorporated into the architecture of the Newark Earthworks. They speculate that this astronomical information is not just symbolically encoded into the site plan, but that the substantial earthen walls, with their long sight lines and a height that corresponds, more or less, to eye level, are massive (and therefore long-lived and tamper proof) fixed instruments for making astronomical observations."[22]

The archaeological picture of the Hopewell-Adena cultures thus presents a compelling fit for a significant portion of the Book of Mormon narrative, particularly the parts having to do with the Land Northward in the later migration periods of both Nephite and Jaredite time periods. It was a thriving, complex civilization, not a desolate wilderness, with capabilities and connections that align re-

---

22. Dr. Brad Lepper as quoted in Joe Knapp, *Hopewell Lunar Astronomy: The Octagon Earthworks*; https://tinyurl.com/5a6enbtm

markably well with the Nephite description, owing most of it to the cultural transmission from Mesoamerica.

Chapter 4

The Zelph Incident: A Key to the Land Northward

The Zelph incident stands as a pivotal historical account, offering a profound geographical anchor for understanding the Book of Mormon's "Land Northward." Despite attempts by some scholars to discredit or reinterpret it, this episode, witnessed by multiple early Church members, directly links Book of Mormon peoples to the mound-building cultures of the Midwestern United States and places the "Land of Desolation" squarely within that region.

On June 3, 1834, during the Zion's Camp expedition, Joseph Smith and his company traveled through Pike County, Illinois. Upon encountering a large Indian mound, they excavated it, uncovering the skeleton of a man. What transpired next was recorded by several eyewitnesses and provides critical geographical insights:

Reuben McBride recorded: "A skeleton was dug up. Joseph [Smith] said his name was Zelph, a great warrior under the Prophet Onandagus. An arrow was found in his ribs . . . which he said he supposed occasioned his death. [He] said he was killed in battle. [He] said he was a man of God, and the curse was taken off or in part. He was a white Lamanite. (He was known from the Atlantic to the Rocky Mountains)."[1]

Moses Martin noted: "This being in the Co[unty] of Pike, here

---

1. See https://tinyurl.com/mr3awzfx ; Ruben McBride Diary, June 3, 1834

we discovered a large quantity of large mounds. Being filled with curiosity, we excavated the top of one some two feet, when we came to the bones of an extraordinary large person or human being; the thigh bones being two inches longer from one socket to the other than of the Prophet, who is upwards of six feet high, which would have constituted some eight or nine feet high. In the trunk of the skeleton near the vitals, we found a large stone arrow which I suppose brought him to his end. Soon after this Joseph had a vision and the Lord shewed him that this man was once a mighty Prophet, and many other things concerning his dead, which had fallen no doubt in some great battles. In addition to this, we found many large fortifications, which also denotes civilization, and an innumerable population, which has fallen by wars and commotion. And the banks of this beautiful river became the deposit of many hundred thousands whose graves and fortifications are overgrown with the sturdy oak, four feet in diameter."[2]

The description of Zelph's skeleton as "extraordinary large" and potentially "eight or nine feet high" aligns somewhat with archaeological findings of "giant" Adena body types in Ohio and elsewhere in the Eastern U.S. Gigantism is well known in all populations throughout the earth.

Wilford Woodruff stated: "While on our travels we visited many of the mounds which were flung up by the ancient inhabitants of this continent, probably by the Nephites and Lamanites. We visited one of those mounds, and several of the brethren dug into it and took from it the bones of a man. Brother Joseph had a vision respecting the person. He said he was a white Lamanite the curse was taken from him, or at least in part. He was killed in battle with an arrow. The arrow was found among his ribs. One of his thigh bones was broken. This was done by a stone flung from a sling in battle years before his death. His name was Zelph. Some of his bones were brought into the camp, and the thigh bone which was broken

---

2. See https://tinyurl.com/5fpm943r ; Moses Martin Journal, June 3, 1834

was put into my wagon, and I carried it to Missouri. Zelph was a large, thick-set man and a man of God. He was a warrior under the great prophet that was known from the hill Cumorah to the Rocky mountains. The above knowledge Joseph received in a vision."[3]

Levi Hancock provided the most geographically explicit statement from Joseph: "On the way to Illinois River where we camped on the west side, in the morning many went to see the big mound about a mile below the crossing. I did not go on it but saw some bones that [were] brought, with a broken arrow.' They [were] laid down by our camp. Joseph addressed himself to Sylvester Smith, 'This is what I told you, and now I want to tell you, that you may know what I meant. This land was called the Land of Desolation, and Onandagus was the King, and a good man was he. There in that mound did he bury his dead. And [they] did not dig holes as the people do now, but they brought there dirt and covered them until, you see, they have raised it to be about one hundred feet high. The last man buried was Zelph. He was a White Lamanite who fought with the people of Onandagus for freedom. When he was young, he was a great warrior, and had his thigh broken, and never was set. It knitted together as you see on the side. He fought after it got strength, until he lost every tooth in his head, save one. When the Lord said he had done enough and suffered him to be killed by that arrow you took from his breast.' These words he said as the camp was moving off the ground. As near as I could learn he had told them something about the mound and got them to go and see for themselves.' I then remembered what he had said a few days before, while passing many mounds on our way that was left of us. Said he, 'There are the bodies of wicked men who have died and are angry at us. If they can take the advantage of us they will, for if we live they will have no hope.' I could not comprehend it but supposed it was all right."[4]

---

3. https://tinyurl.com/2mu3f8vn ; Wilford Woodruff Journal, June 3, 1834
4. https://tinyurl.com/4cyhu225 ; Levi Hancock Journal, June 3, 1834

There can be no doubt that this is a direct quote of Joseph Smith with immense geographical implications giving the one and only direct quote of the prophet, establishing that the area of Illinois comprised part of the Book of Mormon Land of Desolation. There are so many important implications of this statement that it cannot be overstated.

Heber C. Kimball also confirmed: "On Tuesday the 3rd, we went up, several of us, with Joseph Smith Jr. to the top of a mound on the bank of the Illinois river, which was several hundred feet above the river, and from the summit of which, we had a pleasant view of the surrounding country. We could overlook the tops of the trees on to the meadow or prairie on each side the river, as far as our eyes could extend, which was one of the most pleasant scenes I ever beheld. On the top of this mound there was the appearance of three altars, which had been built of stone, one above another, according to the ancient order; and the ground was strewn over with human bones. This caused in us very peculiar feelings, to see the bones of our fellow creatures scattered in this manner, who had been slain in ages past. We felt prompted to dig down into the mound, and send for a shovel and hoe. We proceeded to move away the earth. At about one foot deep we discovered the skeleton of a man, almost entire.' And between two of his ribs we found an Indian arrow, which had evidently been the cause of his death. We took the leg and thigh bones and carried them along with us to Clay County.' All four appeared sound. Elder B[righam] Young has yet the arrow in his possession. It is a common thing to find bones thus drenching upon the earth in this country. The same day, we pursued our journey. While on our way we felt anxious to know who the person was who had been killed by that arrow. It was made known to Joseph that he had been an officer who fell in battle, in the last destruction among the Lamanites, and his name was Zelph. This caused us to rejoice much, to think that God was so mindful of us as to show these things to his servant. Brother Joseph had inquired of the Lord and it was made known in a vision."[5]

The implications of the Zelph incident are profound for Book of Mormon geography:

Hopewell Connection: Zelph and King Onandagus are directly linked to the mound-building cultures of the Midwest, specifically the Hopewell. Joseph Smith's statement that Onandagus was known "from the Hill Cumorah . . . to the Rocky mountains" accurately assesses the geographical scope of the Hopewell Interaction Sphere. This confirms that in Joseph Smith's understanding, Book of Mormon peoples, particularly Nephites and Lamanites, inhabited this expansive North American region, including Illinois and Ohio. This also suggests that there were prophets among the Hopewell nation.

Land of Desolation: Most critically, Levi Hancock's account directly quotes Joseph Smith identifying the Illinois region as "the Land of Desolation." In Book of Mormon internal geography, Desolation is consistently described as the most northern part of the Nephite lands, extending northward from the Narrow Neck of Land.

Narrow Neck Placement: If the Land of Desolation extends into Illinois, then the "Narrow Neck of Land" must necessarily be located *south* of Illinois, specifically in Mesoamerica, according to the internal geography of the Book of Mormon. This direct implication from the most clear and first hand prophetic statement available directly contradicts theories that place the Narrow Neck in the Great Lakes region or anywhere north of Mesoamerica for that matter. This point alone provided a strong impetus, almost the primary impetus in fact, for my shift from a purely U.S.-centric Heartland model to the Dual-Heartland Model. Because if "Joseph knew," as was the Heartlander mantra as early as the time that I was a

---

5. https://tinyurl.com/4k95sspk ; "Extracts from H.C. Kimball's Journal," *Times and Seasons* 6, no. 2 (February 1, 1845):788

heartlander, necessarily in fact, then "Joseph knew." All of the other problematic statements and placements that I used to put stock in suddenly melted away in light of the newer evidence. Now it has been over two decades since I changed my mind on that.

Some scholars have attempted to discredit the Zelph incident in various ways, arguing that its historical accounts contain textual difficulties or later additions, casting doubt on its connection to the Book of Mormon or the identification of Cumorah. Others suggest that "Desolation" in this context might be a descriptive term (meaning "desolate" or "wasted" where a population was destroyed) rather than a proper place name, referring to the Land Northward, primarily inhabited by Jaredites before the time of the Nephites. While a descriptive meaning is plausible in some scriptural contexts (e.g., "desolation of Nehors" for a destroyed city) given Jonathan Neville's approach of multiple working hypotheses, an approach like this may require more mental energy and may result in more cognitive dissonance to try to maintain. When faced with Joseph Smith's explicit statement and the power of its impact, that was exactly what happened to the original Heartland theory in my mind.

The effort by some Mesoamericanist scholars to discredit the implications of this clear statement is understandable, as it directly challenges models that place all Book of Mormon events and Cumorah exclusively in Mesoamerica. However, Joseph Smith's consistent identification of Midwestern mounds and lands as Nephite territory, as seen in his Emma Smith letter, the Adam-Ondi-Ahman statements, and various missionary reports, gives substantial weight to the Zelph incident report in the Hancock Journal as a genuine geographical insight to come to an appropriate understanding of the Geography of the Book of Mormon.

Therefore, the Zelph incident serves as a fundamental geographical anchor. It places the Land of Desolation in a specific region of the Midwestern United States, which, by extension, demands that the Narrow Neck of Land be located further south in Mesoamerica. This insight is critical to the Dual-Heartland Model, providing a bridge between the Mesoamerican heartland and the northern American lands, consistent with both archaeological evidence and prophetic tradition.

Chapter 5

The Narrow Neck of Land and Narrow Pass: The Case for Chivela Pass

My journey in understanding Book of Mormon geography has been one of continuous re-evaluation and refinement. For many years, like a significant number of believing scholars, I held the view that the Isthmus of Tehuantepec in southern Mexico was the most plausible candidate for the "narrow neck of land" mentioned in the Book of Mormon. This perspective was largely influenced by its clear hourglass shape, a seemingly obvious fit for an "isthmus" that connected larger landmasses. My reasoning was based on traditional interpretations that prioritized this particular geographical feature as the primary descriptor.

Some earlier theories, attempting to place all Book of Mormon events within the United States, proposed a "Narrow Neck" in the Great Lakes region, such as the Niagara Peninsula, including my initial proposal that was the first iteration of the modern Heartland Theory that was the prototype that established that as a new paradigm. While the linguistic meaning of "Niagara" ("neck of land," "point of land cut in two," or "bisected bottomland") seemed compelling for a "neck," its geographical fit as the one and only narrow neck mentioned in the Book of Mormon connecting the Land Southward to the Land Northward became problematic in light of other textual and historical evidence, particularly the Zelph incident, as we saw.

However, a deeper dive into the Book of Mormon text, combined with an understanding of ancient Mesoamerican geography and a reconsideration of scholarly interpretations, led me to a significant shift in my thinking, even from the earlier iterations of the Dual-Heartland Theory. I now believe that the "narrow neck of land," "small neck of land" or, more precisely, the "narrow pass" or "narrow passage" (Alma 22:32; Alma 50:34; Alma 63:5; Mormon 2:29; Ether 10:20), is likely the **Chivela Pass** in Mesoamerica, the gap in the Sierra Madre Mountains. This is not merely an intellectual adjustment but a recognition that the textual nuances, when aligned with geological and archaeological realities, point to a different kind of geographical constraint than a simple land bridge between two seas.

The catalyst for this re-evaluation came, in part, as a result of a re-interpretation spawning from my past interpretations even going back to the 1990's exploring alternative interpretations of the "narrow neck" concept, particularly the work of **Richard Hauck**. Hauck, in his writings on Book of Mormon geography, hypothesized that the "narrow neck of land" might not be a slender isthmus surrounded by water on both sides in the traditional sense, but rather a "coastal corridor." This idea suggested that the crucial geographical constriction was a passage *along* a coast, which could still effectively limit movement and create a bottleneck between larger regions. In that case, there was a sea on one side and a different geographic barrier (a mountainous area) on the other. This new way of thinking about the "narrow pass" provided a crucial interpretive tool, allowing for features other than traditional, water-surrounded isthmuses to fit the description.

Richard Hauck's interpretation of a "coastal corridor" gained significant attention within Book of Mormon scholarship. John Clark of Brigham Young University, a prominent Mesoamericanist, reviewed Hauck's work, providing critical analysis of his proposed geographical fits. While Clark's review engaged with the idea of a "coastal corridor" as a legitimate *interpretive approach* to the Book of Mormon text, his assessment did not explicitly endorse or dismiss the feasibility of such a corridor leading to a specific location like Chivela Pass. Rather, Clark's review focused on the general methodological validity of Hauck's interpretations within Mesoamerican geography.[1]

The Chivela Pass, situated in the modern-day Mexican state of Oaxaca, offers a compelling fit for this "narrow pass" concept. Unlike the broad, relatively flat plains of the Isthmus of Tehuantepec, the Chivela Pass is a deep canyon, a significant geological feature that carves a path directly between the two sides of the towering Sierra Madre mountain range. This natural corridor acts as a dramatic bottleneck, funneling travel between the Pacific coast and the Gulf Coast. It is, in essence, a literal "pass" through formidable terrain, much more so than a flat, albeit narrow, strip of land.

The Book of Mormon text, particularly Alma 50:34, lends strong support to my current interpretation: "And it came to pass that they did not head them until they had come to the borders of the land Desolation; and there they did head them, by the narrow pass which led by the sea into the land northward, yea, by the sea, on the west and on the east." My prior assumption, shared by many, was that "by the sea, on the west and on the east" referred to the seas surrounding an isthmus, one reason why I stuck with the Tehuantepec isthmus as the neck for so long, even during the time that I came out with my book, *Resurrecting Cumorah* in the early 2000's when I first published on the earliest version of the Dual-Heartland/Semi-Limited Mesoamerican theory (which this book is essentially an entire rewrite of). However, a re-reading reveals a more precise meaning. The phrase modifies "narrow pass," indicating that the *pass*

---

1. John E. Clark, "A Key for Evaluating Nephite Geographies," Review of Books on the Book of Mormon Vol. 1 (1989), 20-70, particularly discussing Hauck's "coastal corridor" ideas and methodological approach; also see John E. Clark, "Two Points of Book of Mormon Geography: A Review," Review of Books on the Book of Mormon 8/2 (1996), 1-24

itself leads "by the sea" and the area itself is bounded by seas "on the west and on the east." This means the pass, as a geographical feature, lies between or connects areas that are themselves bordered by western and eastern seas. Hauck's coastal corridor that leads up into the area of the Chivela Pass is very likely to be a *part* of a passage, the key part going between the mountain barriers on either side that is actually **called the Narrow Neck of Land**.

The Chivela Pass and the area running along the Pacific coast from the Chivela Pass perfectly embodies all of this. The pass runs North-South, and is bounded by mountains on either side. It provides the most direct and geographically constrained route from the Gulf Coast (eastern side) to the Pacific Coast (western side) in that region of Mesoamerica. It is generally understood by Mesoamericanists that the Pacific is the West Sea, and the Gulf of Mexico is the Sea East. This is something that I agree with. To travel "into the land northward" from that part of the Land Southward, one would naturally traverse this corridor and then up through the Chivela Pass, which then allows for movement to the north. This interpretation removes the need to force the "sea divides the land" phrase to mean a river perhaps flooding an isthmus, as some Tehuantepec advocates have suggested, a point I previously found less convincing, but at the time, didn't have a better suggestion for. The elegance of the Chivela Pass model is that the "sea divides the land" simply means that that is the area where we need to find candidates where the sea makes a division of the land.

One theorist named Theodore Brandley made an observation that seems to lead to an interesting conclusion when applied to the Tehuantepec area. He writes:

> The narrow neck of land was a place where "the sea divides the land," like a major inlet or bay creating a peninsula, rather than an isthmus creating an hour-glass where the land divides

the sea. This narrow neck of land, or peninsula, could be anywhere along a coast-line.[2]

I think his interpretation of the neck of land as a peninsula is wrong. But the idea of a major inlet or bay seems logical in terms of an interpretation of the feature of what the place where the sea divides the land might plausibly be. An inlet of a bay or lagoon makes sense. In fact, "Southward" from the Chivela Pass (or towards the West Sea from the Pass), we see features that fit this description very well, which is a series of coastal lagoons and estuaries on the Pacific side (West Sea side). Some of the major lagoons of this area are: Laguna Superior, Laguna Inferior, Mar Muerto, La Joya-Buenavista (a lagoon system comprising La Joya, Cabeza de Toro, and Buenavista) and Los Patos-Solo Dios (a system of small lagoons comprising Los Patos, El Mosquito, La Balona, Pampita).[3] It appears that a theorist named Rolando Amado, brother to Carlos Amado, an emeritus member of the Seventy, has suggested this area as well for the place where the sea divides the land. In particular, he focused on the inlet for Mar Muerto (apparently the feature called Boca de Tonalá) as a specific division point, in his opinion, which makes sense to me. The author of the "Book of Mormon Resources" Blog, relating the facts about this, states:

> Rolando told me about a powerful experience he had with the text. He thought one of the anchor points he should be able to locate on the modern map is the place where the sea divides the land referenced in Ether 10:20. After years investigating coastal features, he looked at the inlet to Mar Muerto on the coast of Chiapas, Mexico and decided that had to be the place. It was the only feature he had found along the entire coast of

---

2. https://tinyurl.com/amt826nb ; J. Theodore Brandley, Five Misunderstandings Of The Book of Mormon Text That Veils Discovery Of Its Geography, Interpreter Foundation, Dec. 21, 2014

3. https://tinyurl.com/yhrjj452

southern Mexico or Central America that to his mind precisely fit the text.

A person traveling northward along this coast would have to detour 203 kilometers around Mar Muerto, then Laguna Inferior, then Laguna Superior to stay on solid ground and get back to an uninterrupted coastline.

Bro. Amado felt confident this was the place Moroni was describing in Ether chapter 10.[4]

In this region, the sea is dividing the land, literally by way of this series of lagoons and estuaries, where in some cases the fresh water from the land meets and mixes with the water from the sea creating small, brackish water bodies:

> Estuaries are sheltered bodies of water where rivers meet the sea, nutrient-rich freshwater mixes with saltwater, and sunlight penetrates the shallow depths. All of these conditions combine to create some of the most biologically-rich waters on the planet.[5]

Olmec Trade Routes and the Chivela Pass

The archaeological evidence of ancient trade routes in Mesoamerica provides powerful corroboration for the Chivela Pass as a significant geographical feature. The **Olmec civilization**, often identified with the Jaredites (as discussed in chapter 7), was renowned for its extensive and sophisticated trade networks.[6] Their influence, stretching from the Gulf Coast to the Pacific and into the high-

---

4. https://tinyurl.com/ytpdck6d ; "Book of Mormon Resources" Blog, entry for Feb. 8, 2016

5. https://tinyurl.com/5xw8tykn ; National Park Service, Estuaries

6. https://tinyurl.com/mtkxsn7p ; Mark Cartwright, "Olmec Civilization," World History Encyclopedia

lands, necessitated efficient long-distance travel and transportation of goods. Key Olmec trade goods included exotic and valuable materials such as jade, obsidian, serpentine, magnetite, iron ore, and specialized ceramics. These raw materials, often sourced from distant regions, were essential for crafting Olmec ritual objects, tools, and luxury items, indicating robust trade connections across diverse ecological zones.[7]

The **Isthmus of Tehuantepec region**, where the Chivela Pass is located, was an exceptionally critical conduit for Olmec trade. While major Olmec centers like San Lorenzo and La Venta dominated the Gulf Coast, their access to indispensable resources frequently involved traversing this narrow part of the continent. For example, high-quality jade was often sourced from the Motagua Valley in Guatemala, and obsidian from highland sources like El Chayal or San Martin Jilotepeque in the Guatemalan highlands.[8] Transporting these heavy, bulk materials would have been a significant logistical challenge, making the most efficient overland route paramount.

The Chivela Pass, as the most accessible land route through the otherwise rugged Sierra Madre Oriental and Occidental mountains that flank the Isthmus, would have been indispensable for connecting the Olmec heartland on the Gulf lowlands with the Pacific lowlands and the interior highlands. It acts as a natural funnel, concentrating movement through a single, relatively manageable corridor. Archaeological studies have indeed documented the movement of Olmec-style artifacts, raw materials, and cultural ideas across the Isthmus of Tehuantepec. Olmec presence and influence are well-attested on the Pacific slopes of the Isthmus, evidenced by sites like Izapa, which show early monumental art and architectural styles

---

7. https://tinyurl.com/yn6fdeyw ; James Doyle, "Olmec Art," The Met Museum

8. https://tinyurl.com/2327x9e3 ; John E. Clark, Mary E. Pye, and Elizabeth J. Pye, "The Pacific Coast and the Olmec," in *Olmec Art and Archaeology in Mesoamerica*

with clear Olmec connections.[9] This indicates a consistent need for and use of trans-isthmian routes.

Further evidence of Olmec trade and cultural interaction across this narrow zone includes the distribution of unique ceramic styles and sculptural motifs found on both the Gulf and Pacific sides of the Isthmus. The efficiency of movement through the Chivela Pass would have facilitated not only the transport of goods but also the rapid dissemination of Olmec religious and political ideologies, contributing to their pervasive influence across Mesoamerica. The pass would have allowed for relatively quick and secure travel for merchants, diplomats, and even military movements, making it a strategically vital geographical feature for a powerful civilization like the Olmec.

Therefore, the combination of the textual description pointing to a "narrow pass" and the archaeological evidence of extensive Olmec trade traversing the Isthmus of Tehuantepec (for which the Chivela Pass is a prime candidate for the most efficient overland route) strongly supports its identification as the ancient "narrow neck of land" or "narrow pass." This geographical feature would have been the critical choke point for movement between the major population centers of Mesoamerica and the lands to the north, providing a more robust and textually harmonious fit than previous interpretations. The Zelph incident, placing the Land of Desolation in Illinois, necessitates a southern narrow neck in Mesoamerica, and the Chivela Pass stands out as a compelling and archaeologically supported candidate that fits the textual specifics of a "pass" and "neck."

---

9. Gareth W. Lowe, "The Olmec Legacy at Izapa," *Journal of New World Archaeology*, No. 4 (1981), 1-52

Chapter 6

The Olmec as the Jaredites: San Lorenzo and the Narrow Neck

The Book of Mormon introduces us to several ancient peoples who inhabited the American continent. Beyond the Nephites and Lamanites, the text details a much older civilization, the Jaredites, who journeyed to the New World from the Tower of Babel sometime between the 19th and 20th centuries BC. Their history, briefly recounted in the Book of Ether, describes a sophisticated society that flourished for many centuries before ultimately facing a catastrophic destruction. Many scholars, including myself, have found compelling parallels between the Jaredites and the **Olmec civilization** of Mesoamerica, a people who existed in the same general time period and demonstrated many of the advanced cultural traits described in the Book of Mormon. One tradition was recorded by Fray Bernardino de Sahagun, which seems to suggest where the Olmecs landed, at Panuco in the Mexican state of Veracruz:

> The account of the origin of this people, which the old people give, is that they came over the waters from the north. And it is certain that they came in some vessels. It is not known in what manner they were constructed. . . . The first people who came to settle this land came from the direction of Florida, came sailing along the coast, and disembarked in the port of Panuco, which they called Panco and which means, "the place where

those who crossed the waters arrived." This people came in search of the terrestrial paradise, and they brought as watchword 'Tamoanchan,' which means, "We seek our home." And they settled near the highest mountains they found.... It seems that their ancestors possessed some oracle regarding this subject, either from God, or the demon, or from the tradition of the elders, which was handed down to them.[1]

The Olmec civilization, often considered the "mother culture" of Mesoamerica, emerged around 1500 BC in the tropical lowlands of south-central Mexico, particularly in the modern-day states of Veracruz and Tabasco.[2] Their influence, however, extended far beyond this core area, impacting later cultures such as the Maya and Zapotec. The timeline of the Olmec, spanning from approximately 1500 BC to 400 BC, aligns remarkably well with the estimated Jaredite chronology (1900/2000 BC to 300+ BC). This temporal overlap is a fundamental starting point for the proposed correlation. Following is a map of the Olmec heartland:

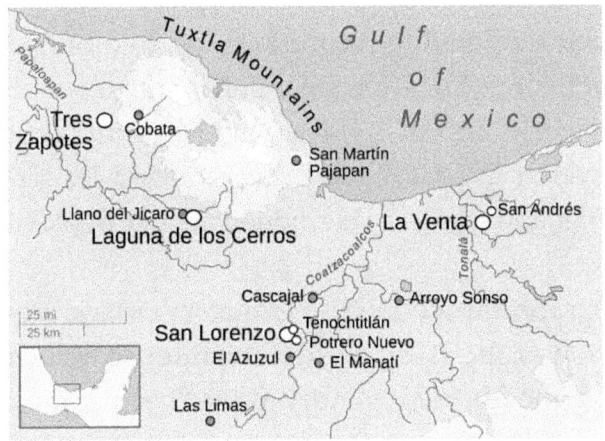

---

1. Bernardino de Sahagun, *Introduction*, in *Florentine Codex: General History of the Things of New Spain*, 13 parts, ed. and trans. Arthur J. O. Anderson and Charles E. Dibble (Santa Fe, NM: School of American Research and the University of Utah, 1982), 1:49, quoted at https://tinyurl.com/26z3kfkj
2. https://tinyurl.com/at4jsw8n ; National Geographic, "Olmec Civilization"

The Olmec possessed many characteristics attributed to the Jaredites. The Book of Mormon describes the Jaredites as a numerous and technologically advanced people capable of monumental construction. They built large cities, engaged in extensive warfare, and developed a complex societal structure. The Olmec, similarly, were known for their impressive monumental architecture, including massive earthen mounds, pyramids, and sophisticated urban planning. Their artistic achievements, particularly the iconic colossal heads carved from basalt boulders, speak to a highly organized society with advanced engineering and artistic capabilities.[3] These colossal heads, some weighing up to 50 tons, were transported many miles from their quarry sites, demonstrating significant logistical prowess and collective labor, akin to the large-scale projects described in Jaredite society.

The Olmec also exhibited signs of a developed political and religious system. Their cities contained elaborate ceremonial centers, and evidence suggests a stratified society with specialized artisans, priests, and rulers. The presence of a sophisticated writing system, represented by early forms of hieroglyphs on stelae and other artifacts, further aligns with the Jaredite record-keeping traditions.[4] While the exact nature of their political organization is still debated, the scale of their public works and the widespread distribution of their cultural influence point to powerful and centralized leadership, much like the kings and leaders mentioned in the Book of Ether.

Crucially for our geographical model, the Olmec heartland lies squarely within the region of the **Isthmus of Tehuantepec**, which includes the Chivela Pass, the feature I identify as the "narrow neck of land" or "narrow pass." This geographical proximity is not co-

---

3. https://tinyurl.com/mtkxsn7p ; Mark Cartwright, "Olmec Civilization," World History Encyclopedia

4. https://tinyurl.com/5n99dytm ; Michael Coe, *Mexico: From the Olmecs to the Aztecs*

incidental. The Book of Mormon text states that the Jaredites built "a great city by the narrow neck of land, by the place where the sea divides the land" (Ether 10:20). This places a significant Jaredite population center directly adjacent to this critical geographical feature. The issue of San Lorenzo being a candidate for this city is one of the key criteria that helps to identify the broader Tehuantepec Isthmus as the area of the narrow neck of land.

One of the most prominent Olmec centers, and a compelling candidate for the "great city" built by Lib near the narrow neck of land, is **San Lorenzo Tenochtitlán**. Its emergence as a key site is often cited as a major criterion for identifying the broader Tehuantepec Isthmus as the region of the narrow neck of land due to its strategic location and early prominence. San Lorenzo was the earliest major Olmec center, flourishing from approximately 1400 BC to 900 BC, making it a fitting candidate for an early Jaredite city.[5] Its timeline aligns with the earlier periods of Jaredite civilization.

San Lorenzo was not merely a collection of dwellings; it was a vast, sprawling complex covering a large plateau, significantly reshaped by human labor. This plateau itself was a monumental construction, involving the movement of enormous quantities of earth to create terraces, platforms, and drainage systems. The city's sheer scale, monumental architecture, and the concentration of early Olmec colossal heads at the site underscore its importance as a major political and economic hub. It controlled a vital floodplain that was highly productive agriculturally, supporting a large population.

The strategic location of San Lorenzo further strengthens its candidacy as the city of Lib. It is situated on the Coatzacoalcos River basin, which provided crucial access to the Gulf Coast. From this location, Olmec influence and trade routes extended across the Isthmus of Tehuantepec. As discussed in the previous chapter, the

---

5. https://tinyurl.com/58dc9mkt; Dumbarton Oaks, "San Lorenzo Tenochtitlán"

Chivela Pass, a deep canyon cutting through the Sierra Madre, served as the most efficient overland route connecting the Gulf Coast to the Pacific Coast and the central highlands. San Lorenzo, as a dominant center on the eastern side of this passage, would have commanded control over a significant portion of the trade and movement flowing through the "narrow pass." The city of Lib, mentioned as being built by a Jaredite king (Lib) who was known for his military prowess and for causing "his people to work sufficiently to bring forth iron from the earth," suggests a powerful and resource-rich center (Ether 10:23). San Lorenzo, with its early monumental feats and strategic location for controlling trade routes, fits this description.

The Book of Mormon also describes the Jaredites as being skilled in metallurgy, working with various metals including iron, copper, and precious metals (Ether 7:9, 10:23). While evidence of extensive metalworking in early Olmec contexts is still a subject of ongoing archaeological investigation, discoveries of iron ore mirrors and other artifacts at Olmec sites indicate their familiarity with and use of metallic substances.[6] If the Olmec are indeed the Jaredites, it suggests that further archaeological discoveries, particularly in relation to their earliest major centers like San Lorenzo, could one day reveal more definitive evidence of their metallurgical practices.

In conclusion, while Jaredite influence definitely extended northward, their core heartland and major cities, including the city of Lib, appear to be firmly established in the Mesoamerican region, as evidenced by sites like San Lorenzo.

The Olmec civilization, with its matching timeline, monumental achievements, complex societal structure, and location within the Isthmus of Tehuantepec region, presents a highly compelling case for being the Jaredite people. The city of San Lorenzo, as the earliest major Olmec center with its strategic location at the confluence of trade routes near the Chivela Pass, emerges as a strong candidate

---

6. https://tinyurl.com/26z3kfkj ; Mirrors in Mesoamerican Culture

for the "great city" built by Lib near the narrow neck of land. While San Lorenzo's prominence solidifies the Tehuantepec Isthmus as the overall region for the narrow neck, my focus on the Chivela Pass as the specific "narrow pass" refines this understanding to a more geographically precise and textually aligned fit. The parallels are not merely coincidental; they provide a consistent framework that aligns the Book of Mormon narrative with the archaeological record of ancient Mesoamerica.

Chapter 7

The River Sidon: Usumacinta vs. Grijalva—Usumacinta Wins

The River Sidon is one of the most frequently mentioned geographical features in the Book of Mormon, playing a critical role in numerous battles and migrations, particularly within the Land Southward. Its characteristics—flowing northward, having a "head" near the city of Manti, and being the site of significant military maneuvers—have made its identification a central point of contention among Book of Mormon geographers. Within Latter-day Saint scholarship, particularly among those who favor a Mesoamerican setting for the Land Southward, the debate has largely centered around two primary candidates: the **Grijalva River** and the **Usumacinta River**.

Scholars in the Church are indeed divided on this issue. For a significant period, the Grijalva River, flowing through the state of Chiapas, Mexico, was a favored candidate. Prominent scholars like the late John Sorenson, while not definitively declaring it, often presented the Grijalva as a strong possibility for the Sidon in his earlier works, primarily due to its northward flow and its proximity to areas that could correspond to Nephite and Lamanite lands.[1] Proponents highlighted its geographical fit with the text's description

---

1. John L. Sorenson, An Ancient American Setting for the Book of Mormon (Salt Lake City: Deseret Book, 1985), 23-24

of the Sidon flowing north, enabling the "down" to Zarahemla and "up" to Nephi directional references.

However, a closer examination of the Book of Mormon text, combined with archaeological and linguistic insights, leads me to conclude that the **Usumacinta River** is the superior candidate for the River Sidon. My reasons for this belief are multi-faceted, drawing on both the textual nuances and compelling external evidence.

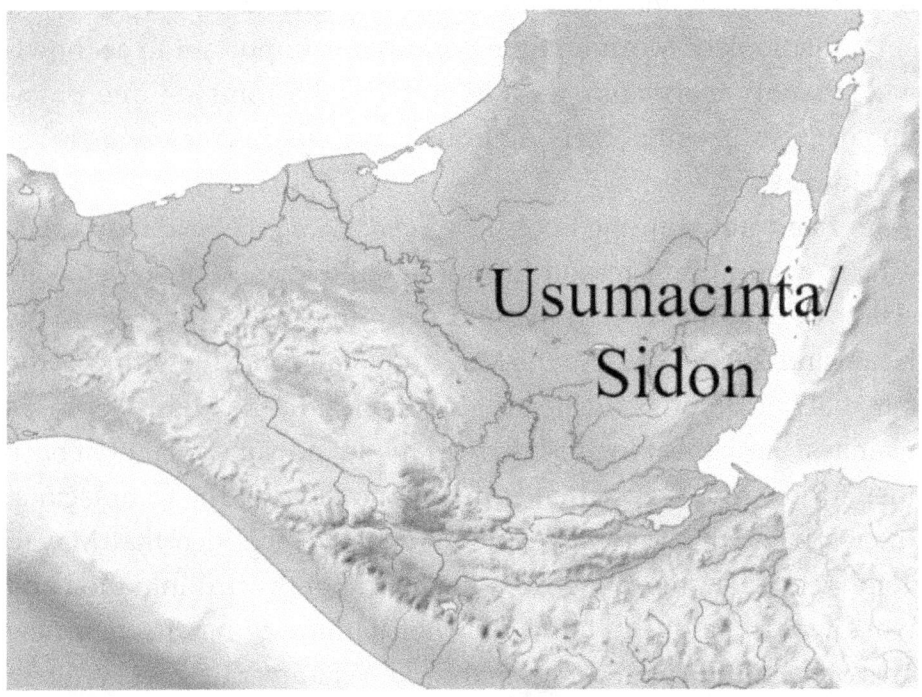

One of the most intriguing arguments for the Usumacinta is a **linguistic connection** to the Hebrew name "Sidon." As previously noted, the Hebrew word "Tsidon" or "Sidon" can mean "fish" or "fishery." John Pratt, notes:

> In addition to the archaeological and geographical evidence, Garth Norman has included in his work some twenty locations where the modern names have roots which can be traced to either Hebrew or Nahuatl words of the same meaning as his proposed Book of Mormon locations.

His list includes examples like "Sidon" being very similar to the Hebrew word for fishery (Tsidon), and that Usumacinta has been deciphered from a Maya glyph to mean "fish river."[2]

This proposed etymology creates a striking parallel, strengthening the case for the Usumacinta as the textual Sidon. It suggests a potential ancient naming convention or recognition of the river's resources that resonated across cultures to ultimately be preserved since the Book of Mormon time period. It is important to acknowledge that other etymologies exist, such as the common interpretation of "Usumacinta" deriving from Mayan words meaning "River of Monkeys."[3]

As I had noted in the book *This Land: Zarahemla and the Nephite Nation*, a significant reason for my initial adoption of some of the claims of Duane Erickson to build upon in the first iteration of the Heartland Theory that Wayne May and I had come up with was the fact that one of the Native American names for the Mississippi was Namaesi Sipu, meaning Fish River. I made the same argument back then for the Mississippi that I am now making for the Usumacinta. Interestingly, this may also lend credence to the idea that Mayan Nephites, upon migrating from the Land Southward into the Land Northward, may have had a habit of naming things in their new area of habitation after things in the old. Whatever the case, William Hamblin wrote the following about Mesoamerican place-names:

> There is no reason to assume that Maya languages, for instance, and Nephite languages were linguistically related. This

---

2. https://tinyurl.com/yc7nk3y7 ; https://tinyurl.com/wyr67pph ; *Meridian Magazine* (12 Dec 2006), "Mormon's Map Puzzle Solved?", John P. Pratt, apparently paraphrasing content from Norman, V. Garth, *Book of Mormon - Mesoamerican Historic Geography* (American Fork, Utah: Ancient America Foundation, 2006), which he quotes elsewhere in the article, and gives a reference to it.

3. Michael D. Coe, *The Maya* (London: Thames & Hudson, 2011), p. 22

further disrupts the continuity of toponyms in the New World.

. . .

>   Pre-Classic Mesoamerican inscriptions . . . [i.e., from Book of Mormon times, the pre-Classic Maya period] are limited to a few dozen. . . . All surviving inscriptional toponyms from Book of Mormon times are . . . symbolic rather than phonetic, making it very difficult, if not impossible, to know how they were pronounced. . . .
>
>   The problem is further complicated by the fact that Mesoamerican toponyms were often translated between languages rather than transliterated phonetically . . . even for those few sites for which a phonetic reading can be determined, the pronunciation of the glyphs seems to have been language-dependent. A Zapotec speaker would pronounce the glyph for the place-name of the same site differently than a Mixtec, and both would be different from Nephite pronunciation, even though all three could theoretically be written with variations of the same glyph.[4]

It seems that Garth Norman's claim seems to be based on the naming of the Usumacinta in some Mayan languages as Xocolha, meaning "waters of the Xoc or Xok," pronounced "Shok." Xoc is a reading of a Maya hieroglyph, meaning fish or water creature. The Maya word itself, not directly connected to the hieroglyph, but used as a *reading* for the hieroglyph, has a regional connotation in some cases more specifically as "shark." On the Book of Mormon Resources blog, the author writes:

>   The Book of Mormon mentions only one river, the Sidon, by name, and it is referenced more than 25 times (e.g. Alma

---

4. https://tinyurl.com/5xajxhka ; Hamblin, William J., *Basic Methodological Problems with the Anti-Mormon Approach to the Geography and Archaeology of the Book of Mormon*

3:3). Almost all references call it "the river Sidon" (e.g. Alma 6:7). There is one reference, though, where Mormon called the Sidon simply "the river" (Alma 43:52). All of these conditions (uniquely named, frequently mentioned, generically named), point to a singular river that dominated its landscape. Most contemporary Book of Mormon modelers (myself included) correlate the Sidon with the mighty Usumacinta, the largest river in Mesoamerica.

I recently found confirmation that the Chontal Maya called the Usumacinta Xocolha which can simply mean "the river." Ronald L. Canter, "Rivers Among the Ruins: The Usumacinta" in PARI Journal, Vol. VII, No. 3, Winter 2007.[5]

Following this author's lead, we look at the reference provided in question:

> Usumacinta, written phonetically as "Usumatsintla" by Teobert Maler (90), is a compound place name formed from the Nahuatl roots osomahtli "monkey," -tzin "small" or "revered," and -tlan "place where X abounds" (Herrera 2004; Karttunnen 983). Thus, osomahtzintla(n) can be literally translated as "Place of Many Sacred (or Small) Monkeys," though it is usually given more broadly as "River of the Sacred Monkey." It was also the name of a Postclassic town on the river near Balancan. Spanish expeditions referred to the upper Usumacinta as the Sacapulas. One source gives the Postclassic name of the Usumacinta above the canyons as Xocolha (Jones 985), and Scholes and Roys (968) give the name as Tanochel at Tenosique. Xocolha means either "Shark River," or simply "The River" in Chontal. A text from Pomona suggests that Pipa' denoted the Usumacinta locally in the Classic. Louis Halle nicknamed the

---

5. https://tinyurl.com/mryj439a ; May 29, 2019, *Book of Mormon Resources* (Blog)

The River Sidon: Usumacinta vs. Grijalva 67

Usumacinta the "River of Ruins" in his book of the same name (94), and that name has also stuck.[6]

As we dig into this more, it becomes clear that the glyph specifically is a water creature or fish, in terms of the identity of the *original glyph* used. This is the **Xoc** glyph:

It just so happens that the Mayan word *Xoc* ("Shoke") that the Maya use as a reading for this glyph, also has a connotation meaning "shark." And we really need to focus on the *identity* of the original *glyph* in question as the generalized water creature, and not so much any particular connotations for the Maya word, even though some have tried to establish a connection between the origin of the English word shark and this Mayan word, because of the similarity in sound. Whether that is the case is not the issue that we are focused on here. We are focused on the generalized meaning of the glyph itself, and its relation to the Usumacinta, notwithstanding any specific regional connotations for the Maya word itself, or any other linguistic puzzles about it. Tim Jones of Humboldt State University writes:

---

6. https://tinyurl.com/38c3z9wx ; Canter, Ronald L., *Rivers Among the Ruins: The Usumacinta*, ThePARIJournal, A quarterly publication of the Pre-Columbian Art Research Institute, Volume VII, No. 3, Winter 2007

In a now classic paper published in 1944, J. Eric S. Thompson attempted to demonstrate the presence of rebus writing among the ancient Maya through the latter's use of a glyph representing an object (a fish) to convey the semantic value of a verb ('to count'). The heart of Thompson's argument was the observation that two Yucatec words, one for 'count' and the other for 'shark,' possessed the same phonetic value: shoke (rhyming with smoke), expressed orthographically as xooc or xoc.[7]

And furthermore, he writes, referring to a certain legendary story about a *xoc* creature that was gathered by the person in his source material from the natives called the "Lacandon," a certain tribe of Maya:

> There is yet another feature of the Lacandon story that merits attention. The river in [the story] was the Usumacinta. But the narrators do not give it that name. They call it Xokla,' which prompts Bruce to interject, "Xokla' appears to be a contraction of Xok-ol-ha,' 'water of the Xok.'" But this Lacandon word is surely a cognate of the Acalan-Chontal xocel haa, recorded in the Maldonado-Paxbolon papers of 1610–12 to describe the water over which Hernan Cortes constructed his famous bridge in 1525, and which Ortwin Smailus has translated as simply 'rio' [river] (Smailus 1975:65). Similarly, the Vienna dictionary offers 'rio' [river] as its translation of the Yucatec x-ocol ha' and x-ocola' (Barerra Vasquez 1980:949). . . .
>
> It appears, then, that the Chontal xocel haa, the Yucatec x-ocol ha' and x-ocola,' and the Lacandon Xokla' are clearly cognates meaning 'river,' even if a particular river in the Lacandon case. But may not Bruce also be correct in his literal reading of

---

7. Jones, Tom, *The Xoc, the Sharke, and the Sea Dogs: An Historical Encounter*, Humboldt State University, Fifth Palenque Round Table, 1983, The Pre-Columbian Art Research Institute, San Francisco, p. 211

Xokla' as 'Water of the Xok'? And if so, then the consequent presence of the concept of the xoc as a river-being in three Maya dialects strongly suggests a fresh water source for the xoc.[8]

He goes on to write:

It is worth noting that the river that enters the Usumacinta from the Mexican side between Piedras Negras and the rapids is called the Rio Chocolja, yet another version, I suggest, of 'water of the xoc.' Similarly, to the west of Lago Izabal, on the Rio Polochic, which empties into the lake (and like it, is notorious for its sharks), there flourished until 1631, a Manche Chol town by the name of Xocolo (Thompson 1 970a:64). That this town, too, should be read 'water of the xoc' receives support from the observation that the immediate upstream neighbors of the Manche Chol were the Pokomchi, the only language group other than Yucatec in which Thompson was able to find the word xoc meaning 'shark.'[9]

Furthermore, the **archaeological and cultural context of Mayan migrations** strongly favors the Usumacinta. Joe Andersen, at one time associated with the BMAF, highlighted that Kaminaljuyu (modern Guatemala City), the regular candidate for the city of Nephi proposed by the Grijalva River advocates, actually dates back to 1500 BC, in the Jaredite time period. It is "too early" to be the city of Nephi, according to his analysis.[10] Nevertheless, because the model I subscribe to in terms of the initial Lehite expansion into Mesoamerica includes a concept also known to others, where I assert that Lehites dominated natives before they mixed in with them.

---

8. Ibid. p. 215-216
9. Ibid. p. 217
10. See Joe Anderson's 2012 Presentation at the BMAF on YouTube: https://tinyurl.com/32upc2by

Since the Book of Mormon was about the Lehites and the house of Israel, they didn't see fit to mention the "others" very much. It was a given among them that the Lehites probably didn't see fit to mention much about the natives eventually just simply mixed in with them, especially after the split between the Nephites and the Lamanites. But Joe Andersen argues for Salama as his candidate for the city of Nephi, rather than Kaminaljuyu. I have no favorite between these two, no dog in that fight, partially because my strategy for Book of Mormon Geography is to try to recognize candidate regions, rather than trying to get so granular as to recognize cities. As readers may have noticed, the one exception to this is my belief that San Lorenzo is the city of Lib because of its nearness to the Chivela Pass as the Narrow Neck/Narrow Pass.

The highland Maya are a compelling fit for the Lamanites, whose territory encompassed the Land of Nephi to the south. Beyond linguistic and migratory patterns, the **Usumacinta River's role as a major ancient trade route** further solidifies its position as the River Sidon. The Usumacinta system, with its extensive network of tributaries, served as a vital transportation artery for the Classic Maya, connecting highland areas with the lowlands and providing access to the Gulf Coast.[11] Goods such as obsidian, jade, cacao, salt, and various foodstuffs moved along this river, facilitating extensive commerce and communication. The river was, in essence, a major "highway" of its time, allowing for rapid movement of people and resources, a characteristic consistent with the Book of Mormon's description of a river central to battles and travel. Certainly it is true that the Grijalva was a route used as well, but the dominance of the Usumacinta as the larger trade route argues for the idea that the local inhabitants viewed it as the more significant and dominant river of the area.

It is important to address some earlier interpretations regarding

---

11. Arthur A. Demarest, *Ancient Maya: The Rise and Fall of a Rainforest Civilization* (Cambridge University Press, 2004), p. 115

the River Sidon's characteristics. Some earlier geographical models, influenced by certain historical statements that placed cities like Manti in Missouri and Zarahemla in Iowa, proposed the Mississippi River as a candidate for the River Sidon. See my initial work on this in the book *This Land: Zarahemla and the Nephite Nation*. These models sometimes interpreted the "head" of the Sidon as a convergence of major tributaries, such as the Missouri River flowing into the Mississippi, rather than its true headwaters. Duane Erickson also interpreted all of the major tributaries as separate heads. While the linguistic argument for "head" having multiple meanings is valid, applying it to the Mississippi and Missouri Rivers to fit a U.S.-centric Sidon becomes incompatible with the overwhelming evidence now pointing to the Usumacinta as the River Sidon within Mesoamerica. Furthermore, earlier proposals (starting with Duane Erickson's) suggested the Sidon flowed southward, out of necessity to make it fit within the US Heartland area. However, this contradicts the seemingly solid interpretation of the internal Book of Mormon text, which I now have come to embrace along with other Mesoamericanists, which clearly indicates the River Sidon flows northward, as peoples went "up" to Nephi (south) and "down" to Zarahemla (north) from the river. In the context of the Usumacinta, the "head of the river Sidon" near Manti would refer to its actual upper reaches within the Mesoamerican highlands, consistent with typical geographical understanding.

Therefore, despite scholarly divisions, the cumulative evidence—linguistic parallels, alignment with crucial Mayan population movements, and its role as a central trade artery—points decisively to the Usumacinta River as the best candidate for the River Sidon.

Chapter 8

Cumorah in New York: Internal Geography and Archaeological Expectations

The Hill Cumorah in Western New York holds a unique and sacred place in Latter-day Saint belief, identified by prophets and early Church leaders as not only the place where Moroni buried the gold plates but also the site of the final, devastating battles of both the Nephite and Jaredite nations. Despite recent Church statements emphasizing the Book of Mormon's doctrinal value over specific geography, and noting that the New York Cumorah "does not readily fit the Book of Mormon description of Cumorah" for some scholars, the weight of historical and traditional testimony, combined with a nuanced understanding of the text, firmly supports its identification. David A. Palmer, a prominent scholar, advanced criteria to disqualify the New York Hill, asserting it could not be the true Cumorah. This stance reflects a common confidence in "proving" geographical points, which I challenge.

Cumorah's Internal Geography:
A Distant Northern Urban Stronghold

Before we get deeper into this, I will give some summaries for the **New York Hill Cumorah arguments for its authenticity.**

The traditional belief that the Hill Cumorah in western New York is the site of the final Book of Mormon battles is primarily based on the statements of early Latter-day Saint leaders and a literal reading of the scriptures. This view gives priority to prophetic authority over external evidence.

Prophetic and Historic Foundation

The core of the argument for this is built on the testimony or conviction about knowledge believed to be had by foundational figures in our faith, who we believe were speaking under divine inspiration on this matter. Furthermore, the Mesoamericanist position contradicts the explicit and consistent teachings of all leaders and prophets since that time, up until recent times, when the Church began to take no position.

Oliver Cowdery's 1835 Letter: In a widely known letter, Cowdery explicitly identified the New York hill as Cumorah, the location where Mormon deposited the sacred records. The fact that Joseph Smith oversaw the letter's publication and never contradicted it is seen as a crucial endorsement.

Joseph Smith and the D&C: While Joseph Smith's early accounts didn't name the hill "Cumorah," later statements attributed to him and a revelation in the Doctrine and Covenants (D&C 128:20) link the coming forth of the Book of Mormon to "Cumorah," reinforcing the New York hill's identity.

The Cave Tradition: Early leaders like Brigham Young spoke of a literal cave inside the hill where Joseph Smith saw numerous ancient records. This is viewed as powerful evidence that the hill was the vast repository described in the Book of Mormon.

Archaeological and Cultural Evidence

Here are factors in support of the New York location.

Sacred Hill: The Seneca Indians, known as Onondawaga or "People of the Great Hill," have a creation story centered on a sacred hill (South Hill) about twenty miles from Palmyra. This demonstrates that sacred hills were significant to Native Americans in the region. Some like me believe that the Hill Cumorah itself served as an "archaeo-astronomical outdoor temple" and an important religious landmark. Its highest prominence looks like a pyramid.

Archaeo-astronomy: In many ancient cultures, the North Star was considered the "omphalos," or the center around which all things turn. Temples and sacred mountains were also viewed as an omphalos. The gathering of the Nephites around the Hill Cumorah with many water sources in the area is compared to King Benjamin's people gathering around the temple. This parallel suggests that Cumorah was treated as a temple.

The Cave and Burial Chambers: It is commonly known that the Hill Cumorah, being a glacial moraine, could not have a natural cave. I bring up the Adena mounds, which were man-made structures similarly composed of sediment, like the New York Hill is. These mounds often contained chambers made from wooden enclosures. I suggest that the same culture could have created a similar man-made structure within the Hill Cumorah to serve as a cave. A pioneer account featured in the movie *17 Miracles* was depicted with a cave and a figure in the movie was presumably an Angel associated with the cave. This seems to be borne out in the historic account. And interestingly, the woman, after being given food by the angel in the cave, went back, and from her perspective, the cave was gone. From the perspective of the Angel, he was looking out at her. While some of these details may not be exactly how it happened, it is telling that a cave can basically disappear from the sight of mortal eyes, and it wasn't even the cave in Cumorah in question. But the principle still seems to be the same. So, some events are left to belief.

The Mesoamericanists have no belief in the New York Hill, so their faith can't help them towards the truth of the matter, because it is a lack thereof. And all they seek to do is to destroy people's faith in the hill and have it in for the hill, which is an unfortunate thing for apologists to do, and is very ironic, since it is usually the mission of apologists to build faith. For New York Cumorah believers, our faith can ultimately help us discover and embrace the truth of it. The idea of two hills named Cumorah is merely a faithless modern apologetic development designed to resolve conflicts their bad interpretations of archaeological findings. The irony is when viewed from the right light, the archaeology is just fine.

The Hypocrisy of the Mesoamericanist Position on Archaeological Evidence at Cumorah Candidate Sites

Critics of the New York Cumorah position often highlight the absence of archaeological evidence—such as large quantities of human remains, weapons, or armor—to prove that the hill near Palmyra was the site of the final Nephite and Jaredite battles. This absence is presented as a fatal flaw in the traditional view. However, a major counter-argument from New York Cumorah proponents is that Mesoamericanists have a similar problem.

No Archaeological Proof for Cerro Vigia: Mesoamericanist scholars have proposed various locations for the Book of Mormon's Cumorah, with **Cerro Vigia** in Veracruz, Mexico, being a popular candidate. This hill is often cited because of its strategic location and proximity to other potential Book of Mormon sites. However, there is no definitive archaeological evidence linking **Cerro Vigia** to the final battles. No vast burial pits, no massive collections of ancient weaponry, and no definitive inscriptions have been found that would prove it was the site of the destruction of the Nephite civilization. It is, in effect, as "archaeologically clean," to use John Clark's words, as the New York Hill.

The Double Standard: This creates a double standard. Critics

(Mesoamericanists) demand a level of archaeological proof for the New York hill that they do not require for their own proposed Mesoamerican sites. They dismiss the New York location on the basis of a lack of evidence, while accepting a Mesoamerican location that also lacks the very evidence they demand. This is a central point of contention and a powerful counter-argument against Mesoamericanist criticisms.

Analysis of the Double Standard

This hypocrisy reveals a fundamental difference in how the two sides approach the Book of Mormon.

Prioritizing External Evidence (Mesoamericanists): The Mesoamericanist position starts with the premise that the Book of Mormon events must be verifiable through external, academic means—namely, archaeology and anthropology. They choose Mesoamerica because the region's archaeology generally aligns with some aspects of the Book of Mormon (e.g., complex civilizations, cities, and warfare). When they can't find direct evidence for a specific site like Cumorah, they argue that the evidence is either yet to be discovered, or the site's identity is still unknown, but the overall framework of a Mesoamerican setting is correct.

Prioritizing Prophetic Evidence (New York Cumorah Proponents): In contrast, the New York Cumorah position starts with the evidence of prophetic statements from Joseph Smith and other early leaders. They see these statements as divinely inspired. When faced with a lack of archaeological proof, they simply point to the same problem down south. They argue that if Mesoamericanists are willing to make excuses for the lack of evidence at **Cerro Vigia**, they should extend the same charitable interpretation to the lack of evidence at the New York hill. But they won't.

Ultimately, the argument about the lack of evidence at **Cerro Vigia** is a strategic point of critique. It highlights that the Mesoamericanist position, for all its supposed academic rigor, is not immune to the same problems it attributes to the traditional, prophetic-based view of the Book of Mormon's geography.

A Consistent Interpretation of Book of Mormon Text

Proponents argue that the New York Cumorah fits a straightforward, literal reading of the Book of Mormon's text, without needing to invent complex geographical models. **Moroni's Final Act:** Mormon 6:6–7 describes Mormon hiding all the records in "the hill Cumorah," and then giving a few plates to his son, Moroni. The text implies a seamless continuity between the battle, the hiding of the records, and Moroni's final act of adding to and burying the plates. A journey of thousands of miles from a Mesoamerican location to New York is seen as an unnecessary and unscriptural addition to this narrative. The most logical interpretation, according to this view, is that Moroni simply completed his work and buried the final record in the same location where his father had hidden the rest.

In conclusion of the summary here, the debate over the location of Cumorah is a microcosm of a larger tension within Latter-day Saint thought: the tension between prophetic authority and external, scientific evidence. The New York Cumorah position is rooted in a tradition of prophetic teachings and a literal reading of foundational texts. I believe that my critiques of Mesoamericanist positions address the scientific and archaeological challenges head-on. The ongoing discussion on these points which seemingly are never really settled by anyone in either camp to the degree to convince or satisfy their opposition, reflects the rich and complex history of a faith seeking to understand and wrestle with its sacred past.

The Book of Mormon provides several key geographical descriptions of Cumorah that, when carefully analyzed, align remarkably well with its New York location within the context of a Dual-Heartland Model:

Exceedingly Great Distance from the Narrow Neck: The text consistently describes Cumorah as being "an exceedingly great distance" from the Land of Zarahemla and the Narrow Neck of Land. For instance, people traveled "many days" to reach the northern lands, indicating a significant journey. This distance is further emphasized by Omer's journey from the Jaredite heartland near the Narrow Neck, traveling "many days" to reach the Hill Shim and the place where the Nephites were destroyed (Cumorah). This explicitly refutes earlier assumptions by some Mesoamericanists that Cumorah had to be in relatively close proximity to the Isthmus of Tehuantepec. The concept of "exceedingly great distance" does not in fact only imply a few hundred, or even a span of a mere 700–1000 miles like the Heartland theory does. In fact, it is well within the realm of plausibility to have a span from Mesoamerica to New York. And with a Hopewellian urban society mostly concentrated in Illinois and Ohio with New York in its backyard, as the satellite heartland where Nephites migrated to later in their history, it makes New York ever more plausible. This distance is indeed "exceedingly great" from a Mesoamerican original heartland.

A Land of Many Waters, Rivers, and Fountains: Mormon explicitly states that Cumorah was located "in a land of many waters, rivers, and fountains." The Great Lakes region of North America, including New York, perfectly fits this description. Minnesota alone is known as "the land of 10,000 lakes," and the region is replete with large bodies of water. Specific local examples near Cumorah include Lake Ontario, Ganargua Creek, Red Creek, and Hathaway Brook, the latter having its source directly from the Hill Cumorah itself. These numerous water sources, including "fountains," are abundant in Western New York. The Book of Mormon's use of "fountain" can refer simply to any river's source, not necessarily bubbling springs as many would insist.

Proximity to a Large Body of Water (Ripliancum): Cumorah is intimately tied to the "waters of Ripliancum, which, by interpretation, is large, or to exceed all" (Ether 15:8). This description aligns with Lake Ontario, which lies just south of the Hill Cumorah in New York. Mormon's strategic choice to gather the Nephite armies at Cumorah, with their backs to a large lake (Mormon 6), demonstrates its military advantage as a natural barrier.

Earlier interpretations in various heartland theory iterations linked Ripliancum to the "Sea West," in some of them with the Great Lakes as the overall "Sea West." This incorrect, since all of the Great Lakes simply qualify as the large bodies of water mentioned in the text (Helaman 3:4)

The Land of Desolation: The Book of Mormon identifies Cumorah as part of the "Land of Desolation," where the Jaredites were destroyed. The Zelph incident (see chapter 5) prophetically confirms that this "Land of Desolation" extended into Illinois, clearly placing it within the Midwestern United States. This means that when the Nephites were driven northward, they were moving towards an area that was known to them as the Land of Desolation. The descriptor "desolate" (Alma 22:30–32) in the text does not necessarily imply a barren desert, but rather a land previously inhabited and most of the inhabitants had been destroyed. And also, it was, coincidentally also without timber due to extensive deforestation by former inhabitants, like the Jaredites. The absence of timber thousands of years ago in the New York area does not preclude it now, as trees would have grown back. The New York area would have been the extreme northward and eastward portion of that land.

Mormon and Moroni's Hopewellian Origins: Mormon himself seems to indicate that he and his father were natives of the Land Northward, specifically near the Hill Shim (which is near Cumorah). John Sorenson writes: "By the time Mormon opens the curtain of history on events in his own lifetime, after A.D. 300, the official Nephite records had long since been moved to the land northward, and he was a native of that area."[1] *His astonishment at*

---

1. Images of Ancient America: Visualizing Book of Mormon Life, p. 202

the immense population density and buildings in the Land Southward[2] *(Mesoamerica) further suggests his origin in a northern region that was also urban, but with the population more spread out, consistent with the archaeology of Hopewell territories. Mormon and Moroni were, in essence, "Hopewellian Nephites," making the New York Cumorah area very likely in the area of their "home" and a logical gathering place for their final stand. Thus, the internal geography indicates a contrast between Mormon's childhood home in the Land Northward compared to the denseness of the large population centers built in the Land Southward.*

Distance to the Eastern Seacoast: While some argue that the New York Cumorah is not close enough to an "eastern seacoast," the Book of Mormon's account of Omer traveling eastward from Cumorah to reach the seashore suggests proximity, even if not immediately on the coast. The Atlantic Ocean is the plausible eastern sea in this context (Ether 9:3)

The sequence of the final Nephite battles, as detailed in Mormon chapters 2–6, shows a prolonged conflict that gradually pushed the Nephites northward from Mesoamerica towards the Hopewellian strongholds and ultimately to Cumorah. This journey involved extensive retreats and re-fortifications through many lands. The fact that they were given four years to gather at Cumorah before the final destruction (Mormon 6:2–5) allowed ample time for people to travel from various parts of the Land Northward, including the Hopewellian centers, to the New York location.

Some earlier U.S.-centric geographical theories proposed that Zarahemla and Manti were also located in the Midwestern United States, citing historical statements placing "Manti" in Missouri and "Zarahemla" in Iowa. These interpretations, while historically significant in demonstrating early Church views of a Nephite presence in the U.S. presence, are incompatible with the broader textual evidence and archaeological correlations in our current model, which places the Land Southward (including Zarahemla and Nephi) in

---

2. Mormon 1:6-7

Mesoamerica. Earlier proposals for the Mississippi River as the Sidon were also tied to this U.S.-centric view, a position that has been superseded by the Usumacinta River in Mesoamerica (see chapter 8).

The Cumorah Criteria Developed by David Palmer to Discount the New York Cumorah

Discussions surrounding the location of the Hill Cumorah often involve lists of criteria used to evaluate potential sites for the final, cataclysmic battles of both nations of the Nephites and Jaredites. One early version of these kinds of lists is found in David Palmer's book *In Search of Cumorah*. His intent on producing this list was certainly to entirely discount the New York Hill as a candidate, not to open-mindedly evaluate what it really has going for it. Variations on this list of criteria are sometimes put forth by those who do not believe the New York location is the authentic site described in the Book of Mormon. The following analysis reviews some of these criteria and demonstrates how the traditional New York Hill Cumorah aligns with them.

Geographic and Environmental Criteria

The geographical features of the Hill Cumorah and its surrounding area in New York are a central part of the argument for its authenticity. Several criteria relate to the hill's location in relation to major land and water features.

Near Eastern Seacoast: One criterion is that the hill must be located near the eastern seacoast. The Hill Cumorah in New York State fits this description, as the state is directly on the East Coast.

Narrow Neck of Land: Another criterion sometimes mentioned is that the hill should be near a "narrow neck of land." This is a bad interpretation of the text of the Book of Mormon, because as we showed, the text speaks of an exceedingly great distance between the neck of land and the Land of many Waters.

Coastal Plain: A similar criterion suggests the hill should be on a coastal plain near mountains and valleys. This, too, is a made-up requirement. It is not truly represented in the Book of Mormon text.

Water Sources and Military Advantage: The Hill Cumorah in New York is said to meet the criteria of being "one day's journey south of a large body of water" and being in an "area of many rivers and waters." Mormon, the Nephite commander, strategically chose an area where his army could position themselves with a large lake at their backs. This natural barrier allowed them to face the enemy from all other sides and provided a military advantage. The text identifies this lake as likely Lake Ontario, or Ripliancum. The area near the hill also provided a major water source for the Nephite nation. In the immediate vicinity, sources like Ganargua Creek and Red Creek are mentioned, with the former having been tapped by the Erie Canal in modern times. And of course, Lake Ontario is a fresh-water source not too far northward

Presence of Fountains: The presence of "fountains" is also listed as a criterion. The Great Lakes region, and the area near the Palmyra hill specifically, contains plenty of water sources that could be described as "fountains." An example is the Hathaway Brook, which flows around the Hill Cumorah and continues along the west side of the Smith family farm. This brook, also known as Crooked Creek or Stafford Creek, has never run dry and its source is on the back side of the Hill Cumorah. The Smith family left the area around the brook uncleared, creating the Sacred Grove where Joseph Smith prayed. The brook was also the site of many early baptisms for prominent members of the Church.

Escape Route: The text also notes that the Great Lakes region offers a "plenty good enough" set of options to satisfy the criterion of having an "escape route southward." Pick anything in the area where you can hide, don't light fires, and don't make noise. But ultimately, as Moroni tells us, they were hunted down and killed eventually anyway.

The Hill's Dimensions and Appearance

The size and prominence of the New York Hill Cumorah are addressed in relation to the battle and its cultural significance.

Viewing the Battlefield: Critics suggest the hill must be "large enough to view hundreds of thousands of bodies" and be a "significant landmark." The Ojibway tradition about a hill in the center of a town on the edge of a lake, from which one could see the entire area, as recorded in the tradition of the destruction of the Mun-dua, supports this. The Hill Cumorah in New York meets this requirement, as Joseph Smith described it as a "hill of considerable size." From the summit, one can see the whole landscape all around, which is consistent with Mormon's account of viewing the fallen people from the hilltop.

Significant Landmark: The Hill Cumorah has an elevation of 117 feet above ground level, making it one of the tallest points in the region. This height relative to the surrounding area is what makes it a notable lookout point. Oliver Cowdery noted that its sudden rise from a plain on the north would attract the notice of travelers. Heber C. Kimball also stated that the Hill Cumorah is a "high hill for that country." The text also suggests that an ancient Jesuit priest's description of a sacred mound where Indians met for worship displays the pattern.

Irrelevant Criteria

Here we are deliberately and explicitly dismissing certain criteria as irrelevant to the Hill Cumorah itself, as they are actually requirements for the "Land Southward" (which is proposed to be Mesoamerica). These criteria include:

Temperate climate with no cold or snow.

A volcanic zone susceptible to earthquakes.

The presence of cities, towers, agriculture, metallurgy, formal political states, and other advanced cultural elements.

While sophisticated cultural traits are associated with the "Land Southward," the sophisticated cultures of the Great Lakes region are considered transplants from Mesoamerica, and therefore do show some of these cultural criteria to a lesser degree.

Archaeological Expectations at Cumorah: Beyond Surface Surveys

For those who adhere to Mesoamerican Cumorah candidates, a frequent argument against the New York Hill is the perceived lack of archaeological evidence for large-scale battles. Critics claim the area is "clean" of artifacts, walls, trenches, or even flint chips. However, such conclusions are often premature, based on limited surface surveys or outdated methodologies.

The absence of abundant surface artifacts does not equate to an absence of a battle. As archaeologists understand, surface surveys offer only "the tip of the iceberg." Many factors, including plowing, erosion, human and animal disturbance, and the depth of buried remains, can obscure or displace artifacts over centuries. A site like Cumorah, which was a "short-term battle site" rather than a "long-term occupation site," would not necessarily be a "chipping area" covered in manufacturing debris; Nephites likely brought their weapons with them.

To ascertain the truth of the archaeological situation at the New York Cumorah, a thorough, professional, and systematic investigation is required:

Subsurface Exploration: Techniques beyond surface surveys are essential. These include augers, corers, shovel test pits, aerial surveys (using cameras, infrared, ground-penetrating radar, thermography), and archaeological geophysics (magnetometers, electrical resistivity). Such methods can detect buried sites, features, and even minute deviations in the Earth's magnetic field caused by artifacts or ancient structures. The discovery of a large stone mass deep within Monks Mound at Cahokia, a Mississippian site in Illinois, illustrates how significant structures can remain hidden beneath the surface, unknown even to experts, until advanced drilling is performed. This demonstrates that the absence of surface evidence does not preclude the existence of substantial buried remains: "During the process of installing horizontal drains to relieve the internal water in Monks Mound that had contributed to several severe slumping episodes along the west side (Second Terrace), the drilling rig encountered stones about 140 feet in and 40 feet below the surface of the Second Terrace. The operator said it felt like 'soft stone,' probably limestone or sandstone, and that it was mostly cobbles or slabs at least six inches in diameter. The drill went through about 32 feet of stones and the drill bit broke off. We have no idea what it is, what shape or size it is, or why it is there. It should not be there. No other cores or excavations have revealed stone in Monks Mound or any other mound at the site, or, as far as we know, at other Mississippian mound sites. We do not know its vertical thickness or the extent of it horizontally, other than the 32 feet that the drill went through."[3] The "experts" had no idea such a stone structure was in Monks Mound until they started drilling, or that such a thing could even exist. It may indicate an earlier stone structure that existed on the site before the mound was built to its current size.

Controlled Excavation: A real archaeological dig under controlled conditions led by credentialed archaeologists in Hopewell/Adena archaeology is necessary. This involves systematically excavating layer by layer, rather than relying on anecdotal reports or unsupervised digs.

Reassessment of Old Data: As John Sorenson notes, it may be less about finding "new" specimens and more about reassessing

---

3. Iseminger, Bill, "Late Breaking News: Stone Mass under Second Terrace of Monks Mound Found," March 1998, https://tinyurl.com/65vv2fps, Cahokia Mounds Museum Society

old ones that have received limited attention by qualified experts. Historical accounts, though not direct archaeological evidence, can serve as valuable guides for setting proper expectations and directing future research. For instance, reports from Sister Susa Young Gates and Elder George Albert Smith from the early 20th century describe farmers finding "basket[s] full" of arrowheads while plowing the Hill Cumorah and surrounding lands, clear evidence of past warfare. While these are "hearsay accounts," they provide a compelling reason for targeted scientific investigation. Likewise, Heber C. Kimball's observation that "The Hill Cumorah . . . had the appearance of a fortification or entrenchment around it" requires follow-up research by archaeologists.

Understanding Cultural Context: Critiques asserting a lack of "impressive" cities or technologies in North America for Book of Mormon times often apply an ethnocentric bias, expecting Mesoamerican-style stone ruins. However, the Hopewell-Adena cultures predominantly built with wood, a material that quickly perishes As Hugh Nibley emphasized, the lack of stone ruins does not imply a lack of advanced civilization or population, and ancient peoples often built with perishable materials. The argument of silence ("we haven't found it, so it didn't exist") is scientifically unsound. The definition of a "city" in the Book of Mormon may also differ from modern academic definitions, referring to scattered settlements around ceremonial centers, which fits the Hopewellian pattern.

It is crucial to remember that the inhabitants involved in the final battles at Cumorah were not solely "New York Natives" in isolation. They were Hopewellian peoples and "Mesoamerican transplants" who traveled to the region from Hopewell cultural centers in Ohio and Illinois, as well as from Mesoamerica itself. Therefore, the archaeology of the local New York tribes is not the sole determinant for evidence; artifacts from the broader Hopewell/Adena sphere should be expected.

The Cumorah Cave and Records: Historical Testimonies

Beyond the battle site, historical accounts from early Church leaders describe a sacred cave within the Hill Cumorah itself, containing vast amounts of ancient records and treasures. These accounts, though not archaeological evidence, add a rich dimension to the site's significance:

Oliver Cowdery and Joseph Smith: Oliver Cowdery purportedly accompanied Joseph Smith into a cave within the Hill Cumorah, describing a room "filled with treasure" and "more plates probably than many wagon loads." Heber C. Kimball corroborated this, describing a vision of a "cave in the hill Cumorah" where "more records than ten men could carry" were "piled up on tables, book upon book." Edward Stevenson further reported that David Whitmer stated Oliver told him this room was "filled with treasure," including the breastplate, the sword of Laban, and untranslated portions of the gold plates. These records were to be kept sacredly until a future time when they would be brought forth.

Divine Protection: The accounts emphasize the divine protection of these records, often implying they were "under the charge of holy angels." The concept of such treasures being "moved from place to place according to the good pleasure of Him who made them and owns them" by priesthood power is suggested. This resonates with ancient traditions, such as the apocryphal book of 2 Baruch, which describes angels hiding the Jerusalem temple's holy vessels in the earth before its destruction, to be restored later (2 Baruch 6:5–8). The Book of Mormon itself supports the idea that God can curse treasures hidden in the earth so that "no man getteth it henceforth and forever" until His appointed time (Helaman 12:18–19). The fact that Cumorah is a glacial moraine (a pile of rocks and dirt) is, in fact, ideal for this type of divine concealment, especially given geological processes like liquefaction during earthquakes that can essentially swallow whole buildings. So, things can be moved about by divine power without any issues whatsoever.

Man-made Chambers: While the Hill Cumorah is a glacial moraine with no *natural* caves, the possibility of man-made chambers within mounds is well-established in Hopewell/Adena archaeology. For instance, the Adena mounds contain large burial chambers or enclosures that contained one or more bodies, often constructed with wooden enclosures and log-lined burial pits. It is entirely plausible that the ancient inhabitants, using their known construction techniques, dug out a part of the hill Cumorah and created a man-made structure in the inside and buried it back up. Accounts from the Grave Creek Mound in West Virginia describe tunnels and burial chambers found within such mounds: "The first person of European descent to discover the mound was early settler Joseph Tomlinson, who literally stumbled off the top while hunting in 1770. In 1838, descendants Jesse and Abelard Tomlinson, and Thomas Briggs gutted the mound, destroying much of the archaeological evidence provided by the scientific study of other mounds. Tunneling from the side and top, the two men discovered a burial chamber in the center containing two skeletons and large amount of jewelry and another room with one skeleton and jewelry. Tomlinson opened the center chamber as a museum, charging 25 cents admission."[4]

Adena mounds contained "large burial chambers or enclosures that contained one or more bodies." The "chambers were surrounded by circular structures with thatched walls and wooden post frames that were burned down, followed by mound construction." And "sometimes canopies were erected over the grave and surrounded by a log platform." Also, "sometimes the burial pit was lined with logs, then roofed over."[5]

Archaeological Digs and Expectations: John Sorenson acknowledges that more research is needed at archaeological sites, and that

---

4. West Virgina Division of Culture and History, https://tinyurl.com/nmas6zj2

5. Native Peoples of North America, https://tinyurl.com/mtxx5kx5

"undistinguished spots" can hold significant evidence. While some archaeological surveys at the New York Cumorah have yielded limited surface artifacts, this does not preclude buried evidence. Factors like plowing, erosion, and looting can severely "scramble" sites, making surface finds unrepresentative of deeper layers. Modern subsurface techniques (radar, magnetometers, and excavations) are needed to truly assess the site's potential. The example of Stonehenge, where extensive digging revealed an entire ancient settlement not visible on the surface, illustrates this point. The New York Cumorah is archaeologically "clean" so-called not because nothing is there, but because insufficient professional, modern archaeological work has been done. John Clark himself, despite critiquing the lack of evidence, is not a specialist in North American archaeology. He has not personally undertaken digs at the New York Cumorah, and his reliance on amateur surface surveys has been critiqued as insufficient for definitive conclusions. The fact that this is what Mesoamericanists rely on for declaring that the site is archaeologically clean shows how they just want to dismiss the site and be quickly done with it.

The Future, Final Destruction of Modern American Societies at Cumorah?

Heber C. Kimball purportedly received a revelation when he prophesied on the destiny of the wicked people of this nation, as reported by his son, J. Golden Kimball: "Heber C. Kimball said it was revealed to him that the last great destruction of the wicked would be on the lakes near the Hill Cumorah."[6] Similar to Heber C. Kimball's statement, it is interesting what Joseph Fielding Smith also stated on this point: "As I stood upon the summit of the Hill Cumorah, in the midst of a vast multitude, only a few of whom

---

6. N. B. Lundwall, *Inspired prophetic warnings to all inhabitants of the earth*, 1940, p. 52

belonged to the Church, I tried to picture the scenes of former days. Here were assembled vast armies filled with bitterness and bent on destruction. I thought of the great promises the Lord had made through his prophets concerning those who should possess this choice land, and how these promises were not fulfilled because the people violated his commandments. Here a people perished because of their extreme wickedness. there must be something in the destiny of things that would cause a repetition of this terrible scene on the same spot many centuries later. I reflected and wondered if this unhappy time would ever come when another still mightier people would incur the wrath of God because of wickedness and likewise perish. If so, would this same spot witness their destruction?"[7] Indeed, hopefully that never happens to our civilization. But if so, the final destruction will not happen to us in Mesoamerica. The spot would be in New York.

These facts, when viewed through the lens of ancient building practices and the principle of divine preservation, reinforce the sacred and historical (and possible future) significance attributed to the New York Hill Cumorah. They suggest that the Hill was not merely a random battleground, but a spot for genocidal wars fought in the same place by divine decree, and also a designated "grand depository" of records, protected and prepared for a future purpose.

---

7. *Doctrines of Salvation*, Vol. III, p. 242

Chapter 9

Other Early Migrations and Ancient Connections

While the Book of Mormon primarily focuses on the Nephite and Lamanite civilizations, it also alludes to other ancient migrations to the Americas. These additional accounts, coupled with archaeological and historical research, suggest a more complex picture of ancient American settlement and interaction. Examining these "other items" provides further context for the broader understanding of the Book of Mormon's historical setting, extending beyond the core Nephite-Lamanite narrative.

The Mulekites and the Phoenician Connection

The Book of Mormon briefly introduces a group of people, whom we refer to as the **Mulekites**, led by Mulek, a surviving son of King Zedekiah of Judah, who fled Jerusalem around 600 BC and arrived in the Americas. Their history is initially distinct from Lehi's group, with whom they eventually merge. This group is noteworthy for its unique origins and potential connections to Old World maritime powers.

Interestingly, several lines of reasoning suggest that many of the Mulekites may have been **Canaanitic Phoenicians**, or at least had significant Phoenician assistance in their transatlantic voyage. This is not a novel suggestion, with scholars like Robert F. Smith and

Hugh Nibley previously noting potential Phoenician ties to Book of Mormon names and voyages.

The strongest evidence for a Phoenician connection lies in the geographical and linguistic clues. The Book of Mormon describes a major river in the Mulekite-settled Land of Zarahemla named **Sidon**. In Hebrew, "Sidon" (Tsidon) can mean "fish" or "fishery." Phoenicia's ancient capitals included Sidon and Tyre, renowned maritime cities. The presence of a Phoenician place-name like Sidon within the core territory of the Mulekites strongly suggests a significant Phoenician influence among them, possibly due to a substantial portion of the group being Phoenician by blood. It is highly improbable that a group primarily composed of Jews would name their principal river after a Phoenician capital unless there was a deep-seated connection.

Furthermore, the **Phoenicians** were arguably the only people in the ancient world with the necessary shipbuilding technology and navigational expertise for a successful transatlantic voyage at that time. They dominated Mediterranean sea travel, trade, and commerce for over two millennia, mastering navigation beyond the "Pillars of Hercules" (Strait of Gibraltar) into the Atlantic. Their ability to circumnavigate Africa around 600 BC (as mentioned by Herodotus) demonstrates their remarkable seafaring capabilities. All of this makes the Phoenicians the ideal candidates to have transported Mulek and his entourage across the Atlantic.

Ancient Greek historian **Diodorus Siculus**, writing around 56 BC, provides a remarkable account that aligns with the Phoenician connection. He describes a large "island" in the "ocean" west of "Libya" (Africa), a voyage of "a number of days to the west." This land, which he describes as fertile, mountainous, with navigable rivers, "costly villas," and "banqueting houses," is clearly a description of America covered by a large population, long before Columbus. Diodorus explicitly states that the Phoenicians discovered this land accidentally after being driven by strong winds far into the ocean. Crucially, he mentions that the Phoenicians of Carthage, when contemplating colonization, decided to keep the existence of this land a secret, intending to use it as a place of refuge "against an incalculable turn of fortune, in case some total disaster should overtake Carthage."[1] This strategic motivation directly parallels Mulek's desperate need for a safe haven unknown to his Babylonian pursuers. It is highly plausible that the Phoenicians offered this secret refuge to their Jewish allies.

The theory that a significant portion of Mulek's group were Phoenicians, speaking other Canaanitic dialects, also provides a compelling explanation for why the Mulekite language became "corrupted" by the time Mosiah's group encountered them. Such linguistic divergence would be natural if the population was a blend of different Semitic-speaking peoples. The Mulekites' landing in the Land Northward (specifically in the Land of Desolation, as the Book of Mormon states they migrated south to Bountiful and Zarahemla) aligns with the interpretation that their arrival point was in Mesoamerica near the Narrow Neck. Some earlier interpretations suggested that the Mulekites landed in the New York area. However, this is incompatible with the evidence placing their subsequent migration southward to Zarahemla, which is in Mesoamerica in our current model.

General Statements on North American Presence

Beyond specific events and peoples, early Church leaders made general statements affirming the Book of Mormon's connection to North America. Joseph Smith referred to the Book of Mormon as giving an account of "the former inhabitants of *this continent*," a phrase which he used in contexts clearly referring to North America. He further explained that the Book of Mormon unfolds the

---

1. https://tinyurl.com/mrxnka7z ; Peter Myers, *Phoenician discovery of America? - Evidence for trans-oceanic contact between ancient civilisations - the Case for Diffusion*, July 9, 2002

history of "ancient America" from its first settlement to the fifth Christian century, describing "two distinct races of people" whose remnant "are the Indians that now inhabit this country." These statements clearly indicated that, in Joseph Smith's understanding, the Book of Mormon events and peoples were strongly tied to the North American continent.

Brigham Young also made specific geographical identifications in Utah:

He identified **St. George, Utah**, as a place where the Gadianton Robbers were found. Some believed this to be in reference to the parts of the Book of Mormon that mention the robbers retreating into the mountains and secret places. This is plausible given the mountainous terrain of the region.

He designated the **St. George Temple site** as a location "dedicated by the Nephites" for a temple, which they were unable to build, a task which the Latter-day Saints were completing for them. Accounts describe the site as initially "boggy" but divinely appointed, with a spring of water that was preserved within the temple's foundation.

He identified the **Manti, Utah Temple site** as a location where Moroni himself stood and dedicated the land for a temple.

These statements, particularly those identifying Nephite activity and dedicated temple sites in Utah, further reinforce the idea that Book of Mormon peoples extended into regions of North America far beyond Mesoamerica. This supports the concept of a broad "Land Northward" that encompassed not just the Midwestern Hopewell areas but also reached into the far western United States, demonstrating the expansive reach of the Nephite nation.

Chapter 10

Indigenous Legends and the Destruction of the "White People"

Beyond archaeological and historical statements, the oral traditions and legends of various Native American tribes in the Eastern United States provide a compelling, albeit "suggestive," layer of corroboration for the Book of Mormon narrative. While not "evidence" in a scientific sense, these myths and stories, as archaeologist Thor Heyerdahl advocated, can serve as "guide books" for historical inquiry, as legends are often rooted in real historical events. My approach here is to treat these traditions seriously as potential echoes of ancient events, leaving it to the reader to draw their own conclusions. It is acknowledged that some may view the use of legendary material as "foolishness," but often "small means . . . confound the wise."

Echoes of the Book of Mormon: Great Wars and Lost Books

Numerous Native American traditions from the Eastern United States recount the destruction of an ancient "white race" in monumental battles, often in locations that align with the Land Northward of our model:

The Art Wakolee Account: Elder Richard Felt recorded a touching account from Art Wakolee, a Sac and Fox Indian from upstate New York. Wakolee recounted traditions passed down from his grandfather about "the great journey of their forefathers across the large waters; of the visit of the great white God, and his promise that he would return; and of the great battle in which many of the Indian people had been killed." Upon hearing the Mormon missionaries, Wakolee recognized the Book of Mormon as the "full story and proof" of his grandfather's legends. The fact that this Native American brother was from New York, and his traditions speak of a great battle, clearly points to events occurring in North America, not Mexico.

The Lenni Lenape (Delaware) Tradition of the Alligewi: John Heckewelder documented a significant tradition from the Delaware Indians, the Lenni Lenape, about a war with the "Talligewi" or "Alligewi." This powerful nation, described as "remarkably tall and stout" with "giants among them," built many "large towns" and "regular fortifications" in the country east of the Mississippi. These fortifications, some near Lake St. Clair and Lake Erie, were described as "walls or banks of earth regularly thrown up, with a deep ditch on the outside." Hundred of slain Alligewi were reportedly buried in large flat mounds outside these entrenchments.

The legend recounts the Lenape and Mengwe (Iroquois) migrating eastward, attempting to cross the Namaesi-Sipu (Mississippi), but being furiously attacked by the numerous Alligewi. After initial retreats, the Lenape allied with the Mengwe and declared war. Great battles ensued, where the Alligewi fortified towns near rivers and lakes. Hundreds fell, buried in heaps. Eventually, the Alligewi, facing inevitable destruction, abandoned the country to the conquerors and fled down the Mississippi River, never to return. The war lasted "many years."

Ojibway (Chippewa) Mun-dua Legend: William W. Warren, an Ojibway historian, recorded a similar tradition of his people annihilating a powerful tribe called the "Mun-dua." This tribe lived in a "single town" so vast that "a person standing on a hill which stood in its centre, could not see the limits of it." This occurred on the "shores of a great lake." The Mun-dua were "fierce and warlike" and ultimately faced annihilation by allied tribes, with women and children perishing in the lake. Their aged chief, a "great medicine man," called upon the "Great Spirit" for help, but it was denied due to their wickedness. A subsequent fog was dispersed by a gale, revealing the fleeing Mun-dua on a hill overlooking their deserted town, leading to their final demise, with a few survivors absorbed into the conquering tribe. This account's details—a vast city up against a great lake, a central hill for viewing, and a wicked people destroyed by an alliance, and a righteous chief—strongly parallel the Nephite destruction at Cumorah.

Kentucky's "White People" and Lost Book: Traditions from tribes northwest of the Ohio River, including the Shawnee and Sac, recount that Kentucky was once settled by a "white people" who possessed "arts unknown to the Indians." These white inhabitants were exterminated in long, bloody wars, with a great battle occurring at the "Falls of the Ohio." After their destruction, "a multitude of human bones were discovered on Sandy Island." Crucially, some of these traditions state that "the Great Spirit had once given the Indians a book which taught them all these arts; but they had lost it and had never since regained a knowledge of them." This detail, particularly the mention of a lost book, is a profound parallel to the Book of Mormon.

New York and Iroquois Traditions: The Iroquois also share similar legends of a "great war" in New York State that lasted "many moons." The Reverend Samuel Kirkland recorded an old, grey-haired Indian describing "thousands . . . slain" and the ground "covered with dead bodies" around forts, with some buried in pits covered with stone and earth. Colonel Joseph Brant, a Mohawk chief, related a tradition of "white men from a foreign country" establishing settlements from the St. Lawrence to the Mississippi, who were later exterminated by an alliance of Indian nations due to jealousy over their numbers and wealth. This tradition directly ties the "white" inhabitants to the mound-building regions and their ultimate destruction in a widespread conflict.

Linguistic Parallels and Physical Traits

Beyond narrative similarities, linguistic and physical descriptions within these legends provide further intriguing parallels:

A Name of a River in the Expected Region of the Land of Zarahemla meaning "Fish River" in a Native Language or Languages of the Area:

As V. Garth Norman has identified, one of the translations of Usumacinta is "Fish River." This is notably similar to the Hebrew word "Tsidon" (Sidon), meaning "fishery" or "to fish." This strong linguistic coincidence suggests that the original meaning of the river's name, likely given by Mulekites, may have been preserved. The concept of writing names with glyphs that represent their meaning (like a fish and a river) means that speakers of different languages would still understand the meaning, even if they pronounced it differently.

"Nephi" in Native Place Names in the Great Lakes Region: The Delaware word "Alligewi" (or "Talligewi"), and the Iroquoian "io" (found in "Ohio" and "Ontario"), mean terms like "most beautiful," "best," "fine," "good," or "large." This parallels the Egyptian word "nefer" (from which "Nephi" is derived), which means "beautiful or fair of appearance," "good," or "goodly of quality." This suggests that the concept of "Nephi" as "beautiful" or "good" was preserved in the names tribes gave to their lands or rivers (e.g., Allegheny as "River of the Alligewi" or "River of the Beautiful People"). The consistency across multiple examples strengthens this linguistic connection.

"Giants" and "Uncommonly Large" People: Many legends, including the Delaware and Shawnee traditions, speak of the ancient "white people" or "Alligewi" as being "remarkably tall and stout" with "giants among them." This aligns with archaeological findings in Ohio (Adena and Hopewell culture) which describe bodies found in mounds as "uncommonly large," often "over six feet tall with powerful builds," some approaching "seven feet." Similar accounts of "giant skeletons" (six to nine feet in length) have been reported from sites across Western New York, Illinois, Minnesota, and Pennsylvania, some in copper armor. These physical descriptions, particularly those validated by archaeological reports of the Adena and Hopewell, offer a "fingerprint" of this lineage's wide geographical spread, including the Zelph skeleton itself. Gigantism is well known across the whole earth. While some may dismiss "giants" as myth, the sheer volume of consistent reports challenges such an easy dismissal.

"Hill People" and Sacred Hills: Iroquoian tribal names like Oneida ("people of the standing stone"), Seneca ("great hill people"), and Onondaga ("people on top of the hills") consistently associate their identity with prominent hills or mounds. Onondaga is clearly a preservation of the name from King Onandagus from the reports about the Zelph incident. This underscores the deep spiritual significance of specific hills, including Cumorah, as religious landmarks, not just geographical features. Accounts from Jesuit priests also describe sacred mounds in the Great Lakes region that were used for worship. The Seneca also revered "South Hill" near Canandaigua Lake, linking it to their creation story.

While these legends offer powerful parallels, it is crucial to approach them with methodological rigor if at all possible. Ethnologists often rely solely on this kind of evidence for hypotheses of migrations and other conclusions. Some scholars are skeptical of using such oral traditions as direct evidence, seeing them as prone to distortion or generic archetypes, or interpolation from outside information not originally in the material. However, legends are often based on real historical events and can be very suggestive. The consistency and specific details across multiple independent tribal accounts defy simple dismissal as mere coincidence or archetype.

While legends may not offer "scientific proof," their "intriguing" parallels warrant serious consideration, especially when attempting to understand the broader historical narrative of a continent for which much ancient detail remains obscure. The sheer number and consistency of these traditions, from various tribes across the Eastern and Midwestern United States, pointing to a great war, a lost white race, and a lost book, provide a compelling cultural echo of the Book of Mormon's account.

The Internal Geography of the Nephite Destruction

A review of the Internal Geography of the Nephite Destruction from the text of the Book of Mormon, from the Narrow Neck of Land at Tehuantepec all the way to Cumorah in New York

First I will establish clearly once again the Geography between the Narrow Neck of Land and Shim/Cumorah area, lest it is forgotten. The hill Shim is the key landmark to this Geography. Shim in the Book of Mormon is mentioned in three places. The first is in relation to when Ammaron was telling Mormon about where to go get the records:

> And about the time that Ammaron hid up the records unto the Lord, he came unto me, (I being about ten years of age, and I began to be learned somewhat after the manner of the learning of my people) and Ammaron said unto me: I perceive that thou art a sober child, and art quick to observe;
>
> Therefore, when ye are about twenty and four years old I would that ye should remember the things that ye have observed concerning this people; and when ye are of that age go to the land Antum, unto a hill which shall be called Shim; and

there have I deposited unto the Lord all the sacred engravings concerning this people.[1]

The next one is in relation to the Nephite flight into the land Northward:

> And now I, Mormon, seeing that the Lamanites were about to overthrow the land, therefore I did go to the hill Shim, and did take up all the records which Ammaron had hid up unto the Lord.[2]

And the next one shows us the relation to Shim with regards to the Narrow Neck of Land (Tehuantepec):

> And the Lord warned Omer in a dream that he should depart out of the land; wherefore Omer departed out of the land with his family, and traveled many days, and came over and passed by the hill of Shim, and came over by the place where the Nephites were destroyed, and from thence eastward, and came to a place which was called Ablom, by the seashore, and there he pitched his tent, and also his sons and his daughters, and all his household, save it were Jared and his family.[3]

So here we see, as discussed previously, that Omer started at the Narrow Neck of Land area where the Jaredite heartland was. And he departed OUT of that land, and then traveled MANY days until he came to Shim. And we see in this verse, that Cumorah is near Shim, being the place where the Nephites was destroyed.

Next, I have already previously established that the hill Cumorah is intimately tied to Ripliancum, being just southward of it, and

---

1. Mormon 1:2-3
2. Mormon 4:23
3. Ether 9: 3

Ripliancum, being one of the Large Bodies of Water in the Land Northward, as we see from Ether 15:8–11, and I quote the verses here again for clarity:

> And it came to pass that he [Coriantumr] came to the waters of Ripliancum, which, by interpretation, is large, or to exceed all; wherefore, when they came to these waters they pitched their tents; and Shiz also pitched his tents near unto them; and therefore on the morrow they did come to battle. And it came to pass that they fought an exceedingly sore battle, in which Coriantumr was wounded again, and he fainted with the loss of blood. And it came to pass that the armies of Coriantumr did press upon the armies of Shiz that they beat them, that they caused them to flee before them; and they did flee southward, and did pitch their tents in a place which was called Ogath. And it came to pass that the army of Coriantumr did pitch their tents by the hill Ramah; and it was that same hill where my father Mormon did hide up the records unto the Lord, which were sacred.

Now I will review all the verses in Mormon with regard to the last battles from the Narrow Neck of Land all the way to Cumorah, but only the verses that have to do with geography. In Mormon chapter 2, the war begins, and the Nephites begin to be driven towards the Great Lakes region, as the first three verses show. Here is verse 1:

> And it came to pass in that same year there began to be a war again between the Nephites and the Lamanites. And notwithstanding I being young, was large in stature; therefore the people of Nephi appointed me that I should be their leader, or the leader of their armies.

In verse 2, this was AD 326:

> Therefore it came to pass that in my sixteenth year I did go forth at the head of an army of the Nephites, against the Lamanites; therefore three hundred and twenty and six years had passed away.

In verse 3, it specifically shows they were going towards the *north countries*:

> And it came to pass that in the three hundred and twenty and seventh year the Lamanites did come upon us with exceedingly great power, insomuch that they did frighten my armies; therefore they would not fight, and they began to retreat towards the north countries.

Verse 4:

> And it came to pass that we did come to the city of Angola, and we did take possession of the city, and make preparations to defend ourselves against the Lamanites. And it came to pass that we did fortify the city with our might; but notwithstanding all our fortifications the Lamanites did come upon us and did drive us out of the city.

Verse 5:

> And they did also drive us forth out of the land of David.

Verse 6:

> And we marched forth and came to the land of Joshua, which was in the borders west by the seashore.

Verse 7 shows that early on, it was Mormon's strategy to try to get all the people together in one:

> And it came to pass that we did gather in our people as fast as it were possible, that we might get them together in one body.

Verse 8:

> But behold, the land was filled with robbers and with Lamanites; and notwithstanding the great destruction which hung over my people, they did not repent of their evil doings; therefore there was blood and carnage spread throughout all the face of the land, both on the part of the Nephites and also on the part of the Lamanites; and it was one complete revolution throughout all the face of the land.

Verse 9 indicates that it was AD 330:

> And now, the Lamanites had a king, and his name was Aaron; and he came against us with an army of forty and four thousand. And behold, I withstood him with forty and two thousand. And it came to pass that I beat him with my army that he fled before me. And behold, all this was done, and three hundred and thirty years had passed away.

Now I skip some verses that don't have to do with geography. But the next geographical verse indicates that a *whole decade and a half had gone in the war* before the Nephites had been driven to the area of Shim, or near the land of Antum in the Hopewellian Cultural Region, as we see in verse 16. This was in AD 345:

> And it came to pass that in the three hundred and forty and fifth year the Nephites did begin to flee before the Lamanites; and they were pursued until they came even to the land of Jashon, before it was possible to stop them in their retreat.

Verse 17 says that Mormon went to the hill Shim for the first time in the Hopewellian region during this first stage of the war:

> And now, the city of Jashon was near the land where Ammaron had deposited the records unto the Lord, that they might not be destroyed. And behold I had gone according to the word of Ammaron, and taken the plates of Nephi, and did make a record according to the words of Ammaron.

As I had shown previously, the other scripture about the hill Shim says that Omer went *out of the land* and took *many days* to get to Shim from the Narrow Neck of Land. This indicates that *Mormon and the rest of the Nephite Nation made this same journey during this first part of the war, just as Omer and his family did*. Verse 20, which is the next verse with geography, shows they went further northward:

> And it came to pass that in this year the people of Nephi again were hunted and driven. And it came to pass that we were driven forth until we had come northward to the land which was called Shem.

Verse 21 again shows Mormon was trying to get everyone gathered in:

> And it came to pass that we did fortify the city of Shem, and we did gather in our people as much as it were possible, that perhaps we might save them from destruction.

Verse 22 shows it was AD 346:

> And it came to pass in the three hundred and forty and sixth year they began to come upon us again.

Now in verses 23–28, Mormon is urging his people to fight more boldly, and they took back their lands, starting with verse 23:

> And it came to pass that I did speak unto my people, and did urge them with great energy, that they would stand boldly before the Lamanites and fight for their wives, and their children, and their houses, and their homes.

Verse 24:

> And my words did arouse them somewhat to vigor, insomuch that they did not flee from before the Lamanites, but did stand with boldness against them.

Verse 25:

> And it came to pass that we did contend with an army of thirty thousand against an army of fifty thousand. And it came to pass that we did stand before them with such firmness that they did flee from before us.

Verse 26:

> And it came to pass that when they had fled we did pursue them with our armies, and did meet them again, and did beat them; nevertheless the strength of the Lord was not with us; yea, we were left to ourselves, that the Spirit of the Lord did not abide in us; therefore we had become weak like unto our brethren.

Verse 27:

> And my heart did sorrow because of this the great calamity of my people, because of their wickedness and their abomina-

tions. But behold, we did go forth against the Lamanites and the robbers of Gadianton, until we had again taken possession of the lands of our inheritance.

Verse 28 shows it was now AD 349, and they had been fighting for three years and had finally triumphed to the point that they had driven the Lamanites out of the Hopewellian lands and all the way down again to the Narrow Neck of Land:

> And the three hundred and forty and ninth year had passed away. And in the three hundred and fiftieth year we made a treaty with the Lamanites and the robbers of Gadianton, in which we did get the lands of our inheritance divided.

Verse 29:

> And the Lamanites did give unto us the land northward, yea, even to the narrow passage which led into the land southward. And we did give unto the Lamanites all the land southward.

So the significance of this review of Mormon chapter 2 is that this first war took the Nephites all the way to the Hopewellian strongholds. But it also took them back down again to the Narrow Neck of Land, in Mesoamerica. So the mistake usually made in interpretation that is made is that the Nephites were driven all the way back to Cumorah for decades. This comes from lack of a careful reading of the text, so it is not true. They had already been fighting a war in the Hopewellian area, and then pushed the Lamanites back down.

Now we go forward with Mormon chapter 3. In this chapter, the Nephites begin in Mesoamerica again. In verse 4, it was AD 360:

> And it came to pass that after this tenth year had passed away, making, in the whole, three hundred and sixty years

from the coming of Christ, the king of the Lamanites sent an epistle unto me, which gave unto me to know that they were preparing to come again to battle against us.

Verse 5:

And it came to pass that I did cause my people that they should gather themselves together at the land Desolation, to a city which was in the borders, by the narrow pass which led into the land southward.

Verse 6:

And there we did place our armies, that we might stop the armies of the Lamanites, that they might not get possession of any of our lands; therefore we did fortify against them with all our force.

In verse 7, it was AD 361, and they were still in Mesoamerica:

And it came to pass that in the three hundred and sixty and first year the Lamanites did come down to the city of Desolation to battle against us; and it came to pass that in that year we did beat them, insomuch that they did return to their own lands again.

In verse 8, they were still in Mesoamerica, and it was AD 362, and they were close enough to the sea to cast their dead into it:

And in the three hundred and sixty and second year they did come down again to battle. And we did beat them again, and did slay a great number of them, and their dead were cast into the sea.

In Mormon 4:1 it was AD 363, and they were still in Mesoamerica:

> And now it came to pass that in the three hundred and sixty and third year the Nephites did go up with their armies to battle against the Lamanites, out of the land Desolation.

Verse 2:

> And it came to pass that the armies of the Nephites were driven back again to the land of Desolation. And while they were yet weary, a fresh army of the Lamanites did come upon them; and they had a sore battle, insomuch that the Lamanites did take possession of the city Desolation, and did slay many of the Nephites, and did take many prisoners.

Verse 3:

> And the remainder did flee and join the inhabitants of the city Teancum. Now the city Teancum lay in the borders by the seashore; and it was also near the city Desolation.

The city Desolation and Teancum were both right by the Narrow Neck of Land at the southern border of Desolation. Now verse 6:

> And it came to pass that the Lamanites did make preparations to come against the city Teancum.

Now for verse 7, it was AD 364, and they were still in Mesoamerica:

> And it came to pass in the three hundred and sixty and fourth year the Lamanites did come against the city Teancum, that they might take possession of the city Teancum also.

Verse 8:

And it came to pass that they were repulsed and driven back by the Nephites. And when the Nephites saw that they had driven the Lamanites they did again boast of their own strength; and they went forth in their own might, and took possession again of the city Desolation.

Verse 9:

And now all these things had been done, and there had been thousands slain on both sides, both the Nephites and the Lamanites.

In verse 10, it was AD 366, and they were still in Mesoamerica:

And it came to pass that the three hundred and sixty and sixth year had passed away, and the Lamanites came again upon the Nephites to battle; and yet the Nephites repented not of the evil they had done, but persisted in their wickedness continually.

Verse 13:

And it came to pass that the Lamanites did take possession of the city Desolation, and this because their number did exceed the number of the Nephites.

In verse 15, it was AD 367, and they were still in Mesoamerica:

And it came to pass that in the three hundred and sixty and seventh year, the Nephites being angry because the Lamanites had sacrificed their women and their children, that they did go against the Lamanites with exceedingly great anger, insomuch that they did beat again the Lamanites, and drive them out of their lands.

In verse 16, there was a break for 8 years while the Lamanites were regrouping to get all of their forces together, and it was AD 375, and they were still in Mesoamerica:

> And the Lamanites did not come again against the Nephites until the three hundred and seventy and fifth year.

Verse 17:

> And in this year they did come down against the Nephites with all their powers; and they were not numbered because of the greatness of their number.

In verse 18, finally the Nephites began to be driven:

> And from this time forth did the Nephites gain no power over the Lamanites, but began to be swept off by them even as a dew before the sun.

In verse 19, it describes how the Nephites lost the city Desolation, and so at this point they started a northward retreat in this last part of the war:

> And it came to pass that the Lamanites did come down against the city Desolation; and there was an exceedingly sore battle fought in the land Desolation, in the which they did beat the Nephites.

In verse 20, the Nephites fled to Boaz, still in Mesoamerica, yet they were headed northward:

And they fled again from before them, and they came to the city Boaz; and there they did stand against the Lamanites with exceeding boldness, insomuch that the Lamanites did not beat them until they had come again the second time.

In verse 21, the Nephites continued to be driven:

And when they had come the second time, the Nephites were driven and slaughtered with an exceedingly great slaughter; their women and their children were again sacrificed unto idols.

Now, here is a key verse in verse 22, because now is the beginning of the actual flight to the Great Lakes, because the Lamanites came at them with such power (see verse 18). The Nephites now begin to abandon all of their towns and villages between Mesoamerica and the Great Lakes:

And it came to pass that the Nephites did again flee from before them, taking all the inhabitants with them, both in towns and villages.

Now in verse 23, Mormon goes ahead of the rest of the Nephites, to fulfill the commandment to take up the records from the Hill Shim, as Ammaron told him to do. Again, I repeat for clarity, as I had shown previously, referring back to the other scripture about the hill Shim, Omer went *out of the land* and took *many days* to get to Shim from the Narrow Neck of Land. This indicates that *Mormon and the rest of the Nephite Nation made this same journey, just as Omer and his family did, this very last time*. Here is verse 23:

And now I, Mormon, seeing that the Lamanites were about to overthrow the land, therefore I did go to the hill Shim, and

Indigenous Legends and the Destruction of the "White People"

did take up all the records which Ammaron had hid up unto the Lord.

Now I move on to Mormon chapter 5. Verses 3 and 4 indicate that the Nephites finally arrived at the Great Lakes Hopewellian strongholds and centers. This shows that the Nephites were indeed making their last stands in the Hopewellian strongholds. The city of Jordan is a Hopewellian city. Here is verse 3:

> And it came to pass that the Lamanites did come against us as we had fled to the city of Jordan; but behold, they were driven back that they did not take the city at that time.

Verse 4:

> And it came to pass that they came against us again, and we did maintain the city. And there were also other cities which were maintained by the Nephites, which strongholds did cut them off that they could not get into the country which lay before us, to destroy the inhabitants of our land.

Verse 5 shows that they got to the Hopewellian strongholds in AD 379, and it gives the detail that they had passed by many lands and to get there:

> But it came to pass that whatsoever lands we had passed by, and the inhabitants thereof were not gathered in, were destroyed by the Lamanites, and their towns, and villages, and cities were burned with fire; and thus three hundred and seventy and nine years passed away.

Verse 6 shows it was AD 380, and the Lamanites are attacking Hopewellian centers:

And it came to pass that in the three hundred and eightieth year the Lamanites did come again against us to battle, and we did stand against them boldly; but it was all in vain, for so great were their numbers that they did tread the people of the Nephites under their feet.

Now in verse 7, it talks about when the Nephites finally gave up trying to maintain the Hopewellian centers.

And it came to pass that we did again take to flight, and those whose flight was swifter than the Lamanites' did escape, and those whose flight did not exceed the Lamanites' were swept down and destroyed.

Now I come to Mormon 6:1:

And now I finish my record concerning the destruction of my people, the Nephites. And it came to pass that we did march forth before the Lamanites.

In verse 2 and 3, finally Mormon secured a temporary treaty with the Lamanite king to gather in all the Nephites:

And I, Mormon, wrote an epistle unto the king of the Lamanites, and desired of him that he would grant unto us that we might gather together our people unto the land of Cumorah, by a hill which was called Cumorah, and there we could give them battle.

Verse 3:

And it came to pass that the king of the Lamanites did grant unto me the thing which I desired.

Indigenous Legends and the Destruction of the "White People" 115

In verse 4, they all traveled to Cumorah:

> And it came to pass that we did march forth to the land of Cumorah, and we did pitch our tents round about the hill Cumorah; and it was in a land of many waters, rivers, and fountains; and here we had hope to gain advantage over the Lamanites.

Now, in verse 5 and 6, it indicates that it had been four years since AD 380 when the Nephites started to lose control of the Hopewellian centers. They all gather at Cumorah:

> And when three hundred and eighty and four years had passed away, we had gathered in all the remainder of our people unto the land of Cumorah.

Next, in verse 6, all the people were finally gathered in and ready for the battle and Mormon buries records in the hill Cumorah, and gives the Book of Mormon plates to Moroni:

> And it came to pass that when we had gathered in all our people in one to the land of Cumorah, behold I, Mormon, began to be old; and knowing it to be the last struggle of my people, and having been commanded of the Lord that I should not suffer the records which had been handed down by our fathers, which were sacred, to fall into the hands of the Lamanites, (for the Lamanites would destroy them) therefore I made this record out of the plates of Nephi, and hid up in the hill Cumorah all the records which had been entrusted to me by the hand of the Lord, save it were these few plates which I gave unto my son Moroni.

And of course, after this, the rest is history, as the Nephites were destroyed. This shows that the Nephite destruction in the Great

Lakes region is perfectly plausible according to this analysis, and shows that these scriptures show that many Nephites occupied the Hopewellian centers. It doesn't prove that all Hopewellians were Nephites, or that all cultures in the Great Lakes region or in other parts of the area that is now the US were Nephites. But it shows that the traditional interpretations are mostly plausible, but with some tweaks on the details and nuances and assumptions that have traditionally been made. The bottom line is, it is easy to take scriptures out of context without contextualizing them.

Mormon's father took him to the Land of Zarahemla from the place in the Land Northward where they resided in the space of only part of one year. Mormon was 10 years old in Mormon 1:2 (about AD 322), and in Mormon 1:3, when he is carried into the Land of Zarahemla by his father, he is 11 years old. Mormon 1:2 notes that it was about the time that Ammaron hid up the records. 4 Nephi 1:48 shows when he hid them up. And Mormon 1:3 shows that it was at the hill Shim where Ammaron hid the plates. Therefore, it is clear that wherever the hill Shim is, Mormon lived nearby, as it was "about" the time that Ammaron had hid up the records (Mormon 1:2). The implication is that the journey taken by Mormon and his father was from the area of the Hill Shim to the Land of Zarahemla. And this journey took something within less than the space of a year's time for them. The journeys of Hagoth's ships into the Land Northward show that travel in the Land Northward was many times by water. William Hamblin states:

> An examination of a map of North America shows that it is possible to sail along the coast of Mexico, up the Mississippi River, and then up the Ohio River to within less than one hundred miles of the New York hill where the plates were buried. Trails and waterways along these major rivers have existed for several thousand years. Sorenson provides a sixteenth-century example of someone walking a similar route in less than a year;

Moroni had thirty-five years between the final battles of the Nephites and when he buried the plates. Thus, the plates could have been transported by canoe to New York, along well-used waterways of the Hopewell Indians (who flourished c. 200 B.C. to A.D. 400).[4]

Mormon 2:20, for example, says that the Nephites were hunted and driven, so the people were fleeing. The people doing that to the Nephites came from all over, from Mesoamerica, and from wherever else anyone had allegiance to those factions. Perhaps there were Lamanite loyalists in the Hopewell areas. So, for example, our country started in New England and the eastern portion of our country in the 1600's where the greatest population density is.

Key Native American Tradition Sources for the Nephite Destruction in the Great Lakes Region

I alluded to some of these and summarized them previously to show the big picture. Now we will give more full versions of the most important of them. The following is more full account of the Allegewi/Tallegwi tradition from (John) Heckewelder in the first volume of the *Transactions of the Historical & Literary Committee of the American Philosophical Society*:

> The Lenni Lenape (according to the traditions handed down to them by their ancestors) resided many hundred years ago in a very distant country in the western part of the American continent. For some reason which I do not find accounted for, they determined on migrating to the eastward, and accordingly set out together in a body. After a very long journey and many nights' encampments by the way, they at length arrived on the Namaesi-Sipu [Mississippi], where they fell in

---

4. *Basic Methodological Problems with the Anti-Mormon Approach to the Geography and Archaeology of the Book of Mormon*, https://tinyurl.com/5xajxhka

with the Mengwe [perhaps the Iroquois], who had likewise emigrated from a distant country, and struck upon this river somewhat higher up. Their object was the same with that of the Delawares; they were proceeding on to the eastward, until they should find a country that pleased them. The spies which the Lenape had sent forward for the purpose of reconnoitering, had long before their arrival discovered that the country east of the Mississippi was inhabited by a very powerful nation who had many large towns built on the great rivers flowing through their land. Those people (as I was told) called themselves Talligew or Tallegewi. Colonel John Gibson, however, a gentleman who has a thorough knowledge of the Indians, and speaks several of their languages, is of opinion that they were not called Tallegewi, but Alligewi, and it would seem that he is right, from the traces of their name, which still remain in the country, the Allegheny river and mountains having indubitably been named after them. The Delawares still call the former Alligewi Sipu, the River of the Alligewi. Many wonderful things are told of this famous people. They are said to have been remarkably tall and stout, and there is a tradition that there were giants among them, people of a much larger size than the tallest of the Lenape. It is related that they had built to themselves regular fortifications or entrenchments, from whence they would sally out, but were generally repulsed. I have seen many of the fortifications said to have been built by them, two of which, in particular, were remarkable. One of them was near the mouth of the river Huron, which empties itself into the Lake St. Clair, on the north side of the lake, at the distance of about 20 miles northeast of Detroit. This spot of ground was, in the year 1776, owned and occupied by a Mr. Tucker. The other works, properly entrenchments, being walls or banks of earth regularly thrown up, with a deep ditch on the outside, were on the Huron River, east of the Sandusky, about six or eight miles from Lake Erie. Outside of the gateway of each of these two entrenchments, which lay within a mile of each other, were a number of large flat mounds in which, the Indian pilot said, were buried hundreds of these slain Alligewi, whom I shall hereafter, with Colonel Gibson, call Alligewi. . . .

When the Lenape arrived on the banks of the Mississippi they sent a message to the Alligewi to request permission to settle themselves in their neighborhood. This was refused them, but they obtained leave to pass through the country and seek a settlement farther eastward. They accordingly began to cross the Namaesi-Sipu, when the Alligewi, seeing that their numbers were so very great, and in fact they consisted of many thousands, made a furious attack upon those who had crossed, threatening them all with destruction, if they dared to persist in coming over to their side of the river. Fired at the treachery of these people, and the great loss of men they had sustained, and besides, not being prepared for a conflict, the Lenape consulted on what was to be done; whether to retreat in the best manner they could, or to try their strength, and let the enemy see that they were not cowards, but men, and too high minded to suffer themselves to be driven off before they had made a trial of their strength and were convinced that the enemy was too powerful for them. The Mengwe, who had hitherto been satisfied with being spectators from a distance, offered to join them, on condition that, after conquering the country, they should be entitled to share it with them; their proposal was accepted, and the resolution was taken by the two nations, to conquer or die.

Having thus united their forces the Lenape and Mengwe declared war against the Alligewi, and great battles were fought in which many warriors fell on both sides. The enemy fortified their large towns and erected fortifications, especially on large rivers and near lakes, where they were successfully attacked and sometimes stormed by the allies. An engagement took place in which hundreds fell, who were afterwards buried in holes or laid together in heaps and covered over with earth. No quarter was given, so that the Alligewi at last, finding that their destruction was inevitable if they persisted in their obstinacy, abandoned the country to the conquerors and fled down the Mississippi River, from whence they never returned.

The war which was carried on with this nation lasted many years, during which the Lenape lost a great number of their warriors, while the Mengwe would always hang back in the rear leaving them to face the enemy. In the end the conquerors divided the country between themselves. The Mengwe made choice of the lands in the vicinity of the great lakes and on their tributary streams, and the Lenape took possession of the country to the south.[5]

R.S. Cotterill, commenting on ancient Kentucky, also commented on the Ancient Allegewi tradition, stating:

There are many traditions to indicate, and a few shreds of evidence to prove, that in the far past that Kentucky supported an advanced and extensive civilization. Nor was it a civilization whose greatness or decline has, like the Romans, left its influence largely written on succeeding ages. It has vanished wholly; the Kentuckians of today owe nothing good or evil to its existence and have no link to connect them with its remains. Yet as this civilization existed on the same soil as we, it becomes the duty, if not the pleasure, of the historian of Kentucky to investigate the remains and describe, if he may, its history.

The Delaware, whom the Indians of every tribe addressed in reverence of their antiquity as "grandfathers" were accus-

---

5. Mercer, H. C., 1885, *The Lenape Stone or The Indian and the Mammoth*, New York & London: G. P. Putnam's Son's, The Knickerbocker Press, Posted at https://tinyurl.com/yzapkfau and https://tinyurl.com/yc6m6vun

tomed to relate as an authentic tradition that eastern North America was at one time occupied by a white people. The Indian name for these was Allegwi. They were no savages or nomads but a nation of fixed habitation and great culture. Whence they had come or when, are points upon which the traditions are silent. But the traditions of the Delaware, the Sac, the Shawnee and even other tribes attest the fact of their presence, their civilization and their power. In the dim past, continue the traditions, the savage Iroquois emerged from the great western country and began to hew their conquering way to the present abode. The Delaware at the same time began migration to the east but took a route much to the south of the Iroquois. Both tribes were confronted and halted on the banks of the Mississippi by the strange Allegewi. But through the Iroquois forced their way restlessly across, the weaker Delaware soon formed an alliance and began a merciless war against their common enemy. The Allegewi in a number of terrific battles were driven southward and finally stood desperately at the bay of their favorite land, Kentucky. Here they built huge mounds for fortifications, for burial places and for temples. How long their last stand respited the Allegewi no one knows, but finally at the falls of the Ohio they staked their lives and fortunes on the issue of one great battle and lost? Their people were expelled and their civilization forgotten.[6]

It seems during the Book of Mormon last battles, something may have happened at the Falls of the Ohio that was a major setback for the Nephites, no doubt one of their last major battles before Cumorah. Or, it may be because of disagreements among medicine men because of conflicts between details in traditions, they couldn't

---

6. R.S. Cotterill, *History of Pioneer Kentucky* as quoted in https://tinyurl.com/mwz2bcpx ; Coryell, Frank and McPhillips, Frank, "Who Were the Talligewi?"

come to agree on exactly where the last battle took place, a fact suggested in William Warren's account that we shall see in a moment.

Brad Lepper, a Hopewell/Adena specialist and trained archaeologist, comments on the Allegewi tradition from the Lenni Lenape, and takes the tradition rather seriously, the way some ethnologists also do:

> We will never know what these people called themselves.... However, some groups have preserved oral traditions which may derive from or relate to the people of the mounds.... The Lenni Lenape, or Delaware, ... met and ultimately vanquished a mighty people who were then the inhabitants of the region. These people were called the Alligewi, or Talligewi, and they are said to have been the builders of the great mounds. The Allegheny River takes its name from these people and, at one time, the Delaware applied that name to the entire Ohio River. Whether or not this tradition describes actual events, it does little to resolve the question of the ethnic identity of the Moundbuilders. However, it would establish they were not Delaware for the Allegewi were here before the Lenni Lenape.[7]

Colonel Joseph Brant (Thayendanega) was a Mohawk chief. His biography was written by William Stone in 1838. In this biography is recorded a conversation between a person named Mr. Woodruff and Brant:

> "Among other things relating to the western country," says Mr. Woodruff, "I was curious to learn in the course of my conversations with Captain Brant, what information he could give me respecting the tumuli [mounds] which are found on and

---

7. Lepper, Bradley T., *People of the Mounds: Ohio's Hopewell Culture*, Ohio Historical Society, Hopewell Culture National Historical Park and Eastern Park and Monument Association, 1995, p. 4-5

near the margin rivers and lakes, from the St. Lawrence to the Mississippi. He stated, in reply, that the subject had been handed down time immemorial, that in an age long gone by, there came white men from a foreign country, and by consent of the Indians established trading-houses and settlements where these tumuli are found. A friendly intercourse was continued for several years; many of the white men brought their wives, and had children born to them; and additions to their numbers were made yearly from their own country. These circumstances at length gave rise to jealousies among the Indians, and fears began to be entertained in regard to the increasing numbers, wealth, and ulterior views of the new comers; apprehending that, becoming strong, they might one day seize upon the country as their own. A secret council, composed of the chiefs of all the different nations from the St. Lawrence to the Mississippi, was therefore convoked; the result of which, after long deliberation, was a resolution that on a certain night designated for that purpose, all of their white neighbors, men, women, and children, should be exterminated. The most profound secrecy was essential to the execution of such a purpose; and such was the fidelity with which the fatal determination was kept, that the conspiracy was successful, and the device carried completely into effect. Not a soul was left to tell the tale."[8]

Here is a Sac and Fox tribal account:

On January 1, 1963, Elder Richard Felt and his companion were invited to dinner at the home of Brother Art Wakolee, a Sac and Fox Indian. Brother Wakolee had joined the Church and served in the presidency of the Cattaraugus Branch in up-

---

8. Trento, Salvatore Michael, 1978, *Search for Lost America: The Mysteries of the Stone Ruins*, Chicago, Illinois: Contemporary Books

state New York. During his conversation with them, he told these two missionaries how, as a child, he had spent much time with his grandfather. He recounted some of the valuable and interesting information he had received from the old man. . . .

Among other things, his grandfather had told him that those sent to teach the truth would always go forth by twos. This was a sign of the true Indian religion. If only one teacher came, the grandfather said to listen to him and pick out that which was good. However, if teachers of religion came in pairs, he should believe all that they said, because it would be true.

The old Indian grandfather had told his grandson some of the ancient traditions of the Indian religion. He told Art about the great journey of their forefathers across the large waters; of the visit of the great white God, and his promise that he would return; and of the great battle in which many of the Indian people had been killed. . . .

After his grandfather had passed on, two teachers of religion came. . . . As they visited him, they told him the story of the book of Mormon; that in it was the story of his forefathers and how they had traveled in boats across the mighty ocean from Jerusalem with the prophet Lehi; of the visit of Jesus Christ to the western continent and of his promise that he would return; and of the final and great battle in which many of the people were killed.

Young Wakolee was very pleased. These Mormon elders had brought him the same stories he had learned as a young boy from his grandfather! He discovered that the ancient Indian religion was basically the same as that taught by The Church of Jesus Christ of Latter-day Saints. . . .

After much study and prayer, Art Wakolee was converted and baptized. He has become a great religious leader among his people. . . .

In the Book of Mormon they are finding the full story and

Indigenous Legends and the Destruction of the "White People"

proof of the legends and traditions that they have believed for so long.[9]

We are told by William W. Warren, an Ojibway Indian, who wrote a history of his people in 1858 that:

> The Waub-ish-a-she, or Marten family, form a numerous body in the tribe, and is one of the leading clans. Tradition says that they are sprung from the remnant captives of a fierce and warlike tribe whom the coalesced Algic tribes have exterminated, and whom they denominate the Mun-dua.[10]

If this is true, the people of that particular family tribe may well preserve Nephite blood. Later in the book, Warren describes the war:

> One tradition, however, is deemed full worthy of notice, and while offering it as an historical fact, it will at the same time answer as a specimen of the mythological character of their tales
> which reach as far back as this period.
> During their residence in the East, the Ojibways have a distinct tradition of having annihilated a tribe whom they denominate Mun-dua. Their old men, whom I have questioned on this subject, do not all agree in the location nor details. Their disagreements, however, are not very material, and I will proceed to give, verbatim, the version of Kah-nin-dum-a-win-so, the old chief of Sandy Lake:
> "There was at one time living on the shores of a great lake,

---

9. Felt, Marie F., "The Great White Father Will Return", Part II, February 1969, *The Instructor*, p. 55-56

10. https://tinyurl.com/ysjk8hve ; Warren, William W., *History of the Ojibway People*, St. Paul, Minnesota: Minnesota Historical Society, 1885, pp. 50

a numerous and powerful tribe of people; they lived congregated in one single town, which was so large that a person standing on a hill which stood in its centre, could not see the limits of it.

"This tribe, whose name was Mun-dua, were fierce and warlike; their hand was against every other tribe, and the captives whom they took in war were burned with fire as offerings to their spirits.

"All the surrounding tribes lived in great fear of them, till their Ojibway brothers called them to council, and sent the wampum and warclub, to collect the warriors of all the tribes with whom they were related. A war party was thus raised, whose line of warriors reached, as they marched in single file, as far as the eye could see. They proceeded against the great town of their common enemy, to put out their fire forever. They surrounded and attacked them from all quarters where their town was not bounded by the lake shore, and though overwhelming in their numbers, yet the Mun-dua had such confidence in their own force and prowess, that on the first day, they sent only their boys to repel the attack. The boys being defeated and driven back, on the second day the young men turned out to beat back their assailants. Still the Ojibways and their allies stood their ground and gradually drove them in, till on the eve of the second day, they found themselves in possession of half the great town. The Mun-duas now became awake to their danger, and on the third day, beginning to consider it a serious business, their old and tried warriors, 'mighty men of valor,' sang their war songs, and putting on their paints and ornaments of battle, they turned out to repel their invaders.

"The fight this day was hand to hand. There is nothing in their traditionary accounts, to equal the fierceness of the struggle described in this battle. The bravest men, probably, in America, had met—one party fighting for vengeance, glory, and renown; and the other for everything dear to man, home, family, for very existence itself!

"The Mun-dua were obliged at last to give way, and hotly pressed by their foes, women and children threw themselves into, and perished in the lake. At this juncture their aged chief, who had witnessed the unavailing defence of his people, and who saw the ground covered with the bodies of his greatest warriors, called with a loud voice on the 'Great Spirit' for help (for besides being chief of the Mun-duas, he was also a great medicine man and juggler).

"Being a wicked people, the Great Spirit did not listen to the prayer of their chief for deliverance. The aged medicine man then called upon the spirits of the water and of the earth, who are the under spirits of the 'Great Spirit of Evil,' and immediately a dark and heavy fog arose from the bosom of the lake, and covered in folds of darkness the site of the vanquished town, and the scene of the bloody battle. The old chieftain by his voice gathered together the remnants of his slaughtered tribe, and under cover of the Evil Spirit's fog, they left their homes forever. The whole day and ensuing night they travelled to escape from their enemies, until a gale of wind, which the medicine men of the Ojibways had asked the Great Spirit to raise, drove away the fog; the surprise of the fleeing Mun-duas was extreme when they found themselves standing on a hill back of their deserted town, and in plain view of their enemies.

"'It is the will of the Great Spirit that we should perish,' exclaimed their old chief; but once more they dragged their wearied limbs in hopeless flight. They ran into an adjacent forest where they buried the women and children in the ground, leaving but a small aperture to enable them to breathe. The men then turned back, and once more they met their pursuing foes in a last mortal combat. They fought stoutly for a while, when again overpowered by numbers, they turned and fled, but in a different direction from the spot where they had secreted their families: but a few men escaped, who afterward returned, and disinterred the women and children. This small remnant of a once powerful tribe were the next year attacked by an Ojibway war-party, taken captive, and incorporated in this tribe. Individuals are pointed out to this day who are of Mun-dua descent, and who are members of the respected family whose totem is the Marten."[11]

John Haywood states in his *History of Tennessee*:

The tradition of the Indians northwest of the Ohio is that Kentucky had been settled by whites and that they had been exterminated by war. They believe that the old fortifications now seen in Kentucky, on the Ohio, were constructed by those white inhabitants. . . . A very aged Shawanese chief on the Auglaise River, concurred in the truth of this tradition. He was 120 years of age, and must have been born some time about the year 1680.

An old Indian informed Mr. Moore that the western country and particularly Kentucky, had been inhabited by white people but that they were destroyed by the Indians; that the last battle was fought at the Falls of the Ohio, and that the Indians drove the aborigines into a small island below the rapids. . . . When the waters of the Ohio had fallen, a multitude of human bones were discovered on Sandy Island, and the Indians told General Clark, of Louisville that the battle of Sandy Island decided finally the fate of Kentucky with its ancient inhabitants. General Clark says, that Kentucke, in the language of the Indians, signifies river of blood. . . . Some Sacs in 1800 told Colonel Joseph Daviess, that Kentucky had been the scene of much blood, and was filled with the manes of its butchered inhabitants. The

---

11. Warren, *History of the Ojibway People*, pp. 91-94

ancient inhabitants, they said, were white and possessed arts unknown to the Indians. Cornstalk told Colonel McKee that it was a current and assured tradition that Kentucky and Ohio had once been settle by white people, possessed of arts not understood by the Indians: that after many severe conflicts they were exterminated. He said the Great Spirit had once given the Indians a book, but the they lost it, and had never since regained the knowledge of the arts. . . . [I]t had been handed down from a very long time ago that there had been a nation of white people inhabiting the country, who made the graves and forts. He said that some Indians who had traveled very far west and northwest, had found a nation of people who lived like Indians, although of a different complexion.[12]

Hugh Nibley verifies for us that some North American Indians have the tradition of the Savior coming to them as do Mesoamerican peoples, contrary to some Mesoamericanist assumptions:

Some years ago I was in Cedar City visiting President Palmer who was a great Indian man. He was a member of the Paiute tribe, who had been initiated, etc. He went out to the place where he had been initiated and told me about the rites, and we went out to the various stations of the place where they do these things, etc. He told me some very interesting things about what happened, the legend and the like. Then, just a week later, I was visiting the Hopis, and they showed me the Hopi Stone. Very few people have seen the Hopi Stone. I was standing out in the dusk. It was getting dark. Mina Lansa was in charge. See, they have a matriarchy, and she is in charge of keeping the sacred records, especially the Hopi stones. There are four of them, and this is the big one. She started saying

---

12. Haywood, John, *Natural and Aboriginal History of Tennessee*, Jackson, Tennessee: McCowat-Mercer Press, 1959, pp. 203-204

"Come here, come here, come here," and it was dark. I thought, what have I done now? These people are very touchy, and I may have offended someone somehow. I went into her house. It was the northernmost house in old Oraibi there, on the mesa. All the elders were sitting around the room, and there was a little kitchen table in the center with an oil lamp on it. She said, "Sit down here." So I sat down at the table. She went into the other room and came back with something wrapped in a blanket. She unwrapped it, and that was the Hopi Stone that very few people get to see. It was beautiful, porphyry—heavy, so big, and an inch and a half thick, highly polished, covered completely with characters on both sides. I recognized immediately what the main theme was, and I started to talk to them about it. It showed the people holding hands. I had learned this from President Palmer just a week before; this helped me out. The people were wicked, and there was a great destruction, a great earthquake, and terrible things happened. The people were frightened, and they were totally in the dark. They didn't know what to do, so a voice came and told them all to hold hands. So they all held hands, and then they heard a voice above them. They looked up to heaven and they saw a little point of light coming. It got brighter and brighter and brighter, and a man came down. It was Mashiach, the Messiah. I started to tell them this, and Mina Lansa grabbed the rock out of my hands. She said "You're a smart man. You know a lot, but you don't know everything." She wrapped it up and took it out. She wasn't going to hear any more. But they recognized it, and it caused a great hum to go on, etc. So they have this legend about the Savior who came from above (the Southwest Indians still have it) and he descended while the people were waiting in the blackness. They could see him, and he came down and taught them.[13]

---

13. Hugh Nibley, *Teachings of the Book of Mormon, Semester 3*, pp. 328-329)

Key LDS Historical Statements Relevant to Geography and Cumorah

As we reviewed earlier, Levi Hancock recorded Joseph Smith as saying:

> This land was called the Land of Desolation, and Onandagus was the King, and a good man was he. There in that mound did he bury his dead."[14]

As we noted, this portion is possibly the most important statement on geography out of all of them, because it establishes from Joseph Smith's own mouth the identity of the land of the area, that it was part of the Land of Desolation, known to be part of the domain previously inhabited in the Land Northward by the Jaredites. Therefore, from this evidence alone, the area cannot be the Land of Zarahemla, as suggested by the Heartland model. The core of the Heartland Model claim is that "Joseph knew." Well, what did he know? This. This one statement caused me ultimately to change my mind entirely on Book of Mormon Geography, because it is the strongest evidence of all, about what "Joseph knew." If I am being intellectually honest, which I try at all times to be, I simply cannot and could not ignore its implications. I'm not saying at all that people that think it is merely a descriptor of the area are not being honest, because that is the way they deal with the evidence. I just think I have to be honest with myself that it ultimately has never sat well with me to try to explain this one away, and not simply let the statement speak in plainness. There is nothing that compels me to explain this away, so I would rather not, and let the implications fall where they may, which is exactly what happened. D&C 128:29–20 says:

> And again, what do we hear? Glad tidings from Cumorah!

---

14. https://tinyurl.com/4cyhu225 ; Levi Hancock Journal, June 3, 1834

Moroni, an angel from heaven, declaring the fulfillment of the prophets—the book to be revealed.

John Sorenson said the following about this verse, trying to claim that Joseph Smith and the people generally "by that time," were calling it Cumorah, to somehow distinguish that from an earlier time period, as if Joseph Smith *wasn't* calling it Cumorah earlier than that, which is absurd:

> It is clear that by the date of this revelation, Joseph Smith, and seemingly his readers generally, commonly recognized the term Cumorah to refer to the hill in New York.[15]

Contradicting Sorenson, Lucy Mack Smith, the mother of the Prophet wrote that Joseph Smith referred to the hill near Palmyra as "Cumorah" in such an early time period, that it is unmistakably clear that it was the time before the Book of Mormon came forth, from the context. Here she quotes the Prophet:

> "Stop, father, stop," said Joseph, "it was the angel of the Lord. As I passed by the hill of Cumorah, where the plates are, the angel met me and said that I had not been engaged enough in the work of the Lord; that the time had come for the record to be brought forth; and that I must be up and doing and set myself about the things which God had commanded me to do. But, father, give yourself no uneasiness concerning the reprimand which I have received, for I now know the course that I am to pursue, so all will be well.[16]

Here, according to his mother, we are seeing Joseph Smith say

---

15. Sorenson, John, *The Geography of Book of Mormon Events: A Source Book*, Provo, UT: FARMS, 1992, p. 374

16. Smith, Lucy Mack, 1954, *History of the Prophet Joseph Smith*, Salt Lake City, UT: Bookcraft, p. 100

that the hill in New York is Cumorah, without any hesitation, and without trying to make up some other hill far to the south. Next, is part of Oliver Cowdery's speech to the Delaware Indians, in which he unmistakably refers to the hill called by Moroni himself the Hill Cumorah, the very hill in New York:

> Once the red men were many; they occupied the country from sea to sea—from the rising to the setting sun; the whole land. . . .
>
> Thousands of moons ago, when the red men's forefathers dwelt in peace and possessed this whole land the Great spirit talked with them, and revealed His law and His will and much knowledge to their wise men and prophets. This they wrote in a Book . . . written on plates of gold and handed down from father to son for many ages and generations.
>
> It was then that the people prospered and were strong and mighty; they cultivated the earth, built buildings and cities and abounded in all good things, as the pale faces now do. . . .
>
> This Book, which contained these things was hid in the earth by Moroni, in a hill called by him Cumorah, which hill is now in the state of New York, near the village of Palmyra, in Ontario county.[17]

Oliver Cowdery also stated, in a letter to W. W. Phelps (numbered letter 8), published in the Messenger and Advocate:

> The hill of which I have been speaking, at the time mentioned, presented a varied appearance: the north end rose suddenly from the plain, forming a promontory without timber, but covered with grass. As you passed to the south you soon came to scattering timber, the surface having been cleared by

---

17. Autobiography of Parley P. Pratt, pp. 56-61; Documentary History of the Church Vol 1: Footnotes 183:2-18).

art or by wind; and a short distance further left, you are surrounded with the common forest of the country. It is necessary to observe, that even the part cleared was only occupied for pasturage, its steep ascent and narrow summit not admitting the plow of the husbandman, with any degree of ease or profit. It was at the second mentioned place where the record was found to be deposited, on the west side of the hill, not far from the top down its side; and when myself visited the place in the year 1830, there were several trees standing: enough to cause a shade in summer, but not so much as to prevent the surface being covered with grass-which was also the case when the record was first found. . . .

How far below the surface these records were placed by Moroni, I am unable to say; but from the fact that they had been some fourteen hundred years buried, and that too on the side of a hill so steep, one is ready to conclude that they were some feel below, as the earth would naturally wear more or less in that length of time. But they being placed toward the top of the hill, the ground would not remove as much as at two thirds, perhaps. Another circumstance would prevent a wearing of the earth: in all probability, as soon as timber had time to grow, the hill was covered, after the Nephites were destroyed, and the roots of the same would hold the surface. However, on this point I shall leave every man to draw his own conclusion, and form his own speculation, as I only promised to give a description of the place at the time the records were found in 1823.- It is sufficient for my present purpose, to know, that such is the fact: that in 1823, yes, 1823, a man with whom I have had the most intimate and personal acquaintance, for almost seven years actually discovered by the vision of God, the plates from which the book of Mormon, as much as it is disbelieved, was translated! Such is the case, though men rack their very brains to invent falsehoods, and then waft them upon every breeze, to the contrary notwithstanding.

I have now given sufficient on the subject of the hill Cumorah—it has a singular and imposing appearance for that country, and must excite the curious inquiry of every lover of the book of Mormon. . . . The manner in which the plates were deposited:

First, a hole of sufficient depth, (how deep I know not,) was dug. At the bottom of this was laid a stone of suitable size, the upper surface being smooth. At each edge was placed a large quantity of cement, and into this cement, at the four edges of this stone, were placed, erect, four others, their bottom edges resting in the cement at the outer edges of the first stone. The four last named, when placed erect, formed a box, the corners, or where the edges of the four came in contact, were also cemented so firmly that the moisture from without was prevented from entering. It is to be observed, also, that the inner surface of the four erect, or side stones was smooth. This box was sufficiently large to admit a breast plate. . . . From the bottom of the box, or from the breast-plate, arose three small pillars composed of the same description of cement used on the edges; and upon these three pillars was placed the record of the children of Joseph, and of a people who left the tower far, far before the days of Joseph, or a sketch of each. . . .

[T]his box, containing the record was covered with another stone, the bottom surface being flat and the upper, crowning. But those three pillars were not so lengthy as to cause the plates and the crowning stone to come in contact. I have now given you, according to my promise, the manner in which this record was deposited; though when it was first visited by our brother, in 1823, a part of the crowning stone was visible above the surface while the edges were concealed by the soil and grass.[18]

Here, he describes that he knew the Prophet Joseph with such an amount of intimacy for seven years, that it makes no sense whatsoever to discount the facts on who he got the information from about the name and identity of the hill. But in Letter 7 of Cowdery to Phelps, he really gets specific, in such plainness that no man can deny his intent. They can deny the accuracy of his information, as they continue do, but no man can deny the meaning of the words or twist them:

> I must now give you some description of the place where, and the manner in which these records were deposited.
>
> You are acquainted with the mail road from Palmyra, Wayne Co. to Canandaigua, Ontario Co. N. Y. and also, as you pass from the former to the latter place, before arriving at the little village of Manchester, say from three to four, or about four miles from Palmyra, you pass a large hill on the east side of the road. Why I say large, is, because it is as large perhaps, as any in that country. To a person acquainted with this road, a description would be unnecessary, as it is the largest and rises the highest of any on that route. The north end rises quite sudden until it assumes a level with the more southerly extremity, and I think I may say an elevation higher than at the south a short distance, say half or three fourths of a mile. As you pass toward Canandaigua it lessens gradually until the surface assumes its common level, or is broken by other smaller hills or ridges, water courses and ravines. I think I am justified in saying that this is the highest hill for some distance round, and I am certain that its appearance, as it rises so suddenly from a

---

18. *Latter-day Saints Messenger and Advocate*, Vol. 2 No. 1, pp. 37-39 (Letter 8), emphasis added, spelling corrected

plain on the north, must attract the notice of the traveler as he passes by.

At about one mile west rises another ridge of less height, running parallel with the former, leaving a beautiful vale between. The soil is of the first quality for the country, and under a state of cultivation, which gives a prospect at once imposing, when one reflects on the fact, that here, between these hills, the entire power and national strength of both the Jaredites and Nephites were destroyed.

By turning to the 529th and 530th pages of the Book of Mormon, you will read Mormon's account of the last great struggle of his people, as they were encamped round this hill Cumorah. (It is printed Camorah [i.e. in the earliest version of the text], which is an error.) In this valley fell the remaining strength and pride of a once powerful people, the Nephites—once so highly favored of the Lord, but at that time in darkness, doomed to suffer extermination by the hand of their barbarous and uncivilized brethren. From the top of this hill, Mormon, with a few others, after the battle, gazed with horror upon the mangled remains of those who, the day before, were filled with anxiety, hope, or doubt. A few had fled to the South, who were hunted down by the victorious party, and all who would not deny the Savior and his religion, were put to death. Mormon himself, according to the record of his son Moroni, was also slain.

But a long time previous to this national disaster it appears from his own account, he foresaw approaching destruction. In fact, if he perused the records of his fathers, which were in his possession, he could have learned that such would be the case. Alma, who lived before the coming of the Messiah, prophesies this. He however, by Divine appointment, abridged from those records, in his own style and language, a short account of the more important and prominent items, from the days of Lehi to his own time, after which he deposited, as he says, on the 529th page, all the records in this same hill, Cumorah, and after gave his small record to his son Moroni, who, as appears from the same, finished it, after witnessing the extinction of his people as a nation.

It was not the wicked who overcame the righteous: far from this: it was the wicked against the wicked, and by the wicked the wicked were punished. The Nephites who were once enlightened, had fallen from a more elevated standing as to favor and privilege before the Lord, in consequence of the righteousness of their fathers, and now falling below, for such was actually the case, were suffered to be overcome, and the land was left to the possession of the red men, who were without intelligence, only in the affairs of their wars; and having no records, only preserving their history by tradition from father to son, lost the account of their true origin, and wandered from river to river, from hill to hill, from mountain to mountain, and from sea to sea, till the land was again peopled, in a measure, by a rude, wild, revengeful, warlike and barbarous race. Such are our Indians.

This hill, by the Jaredites, was called Ramah: by it, or around it, pitched the famous army of Coriantumr their tent. Coriantumr was the last king of the Jaredites. The opposing army were to the west, and in this same valley, and near by. From day to day, did that mighty race spill their blood, in wrath, contending as it were, brother against brother, and father against son. In this same spot, in full view from the top of this same hill, one may gaze with astonishment upon the ground which was twice covered with the dead and dying of our fellowmen. Here may be seen, where once sunk to naught the pride and strength of two mighty nations; and here may be contemplated in solitude, while nothing but the faithful record of Mormon and Moroni is now extant to inform us of the fact, scenes of misery and distress—the aged, whose silver locks in other places, and at other times, would command reverence; the mother, who, in other circumstances would be spared from violence—the infant, whose tender cries would be regarded and listened to with a feeling of compassion and tenderness—and the virgin, whose grace, beauty and modesty, would be esteemed and held inviolate by all good men and enlightened and civilized nations, were alike disregarded and treated with scorn! In vain did the hoary head and man of gray hairs ask for mercy—in vain did the mother plead for compassion—in vain did the helpless and harmless infant weep for very anguish—and in vain did the virgin seek to escape the ruthless hand of revengeful foes and demons in human form—all alike were trampled down by the feet of the strong, and crushed beneath the rage of battle and war! Alas! who can reflect upon the last struggles of great and populous nations, sinking to dust beneath the hand of justice and retribution, without weeping over the corruption of the human heart, and sighing for the hour when the clangor of arms shall no more be heard, nor the calamities of contending armies be any more experienced for a thousand years? Alas! the calamity of war, the extinction of nations, the ruin of kingdoms, the fall of empires, and the dissolution of governments! Oh! the misery, distress and evil attendant, on these. Who can contemplate like scenes without sorrowing, and who so destitute of commiseration as not to be pained that man has fallen so low, so far beneath the station in which he was created?

In this vale lie commingled, in one mass of ruin, the ashes of thousands, and in this vale were destined to be consumed the fair forms and vigorous systems of tens of thousands of the human race—blood mixed with blood, flesh with flesh, bones with bones, and dust with dust! When the vital spark which animated their clay had fled, each lifeless lump lay on one common level—cold and inanimate. Those bosoms which had burned with rage against each other for real or supposed injury, had now ceased to heave with malice; those arms which were a few moments before nerved with strength, had alike become paralyzed, and those hearts which had been fired with revenge, had now ceased to heave with malice; those arms which were a few moments before nerved with strength, had alike become paralyzed, and those hearts which had been fired with revenge, had now ceased to beat, and the head to think—in silence, in solitude, and in disgrace alike, they have long since turned to earth, to their mother dust, to await the august, and to millions, awful hour, when the trump of the Son of God shall echo and re-echo from the skies, and they come forth quickened and immortalized, to not only stand in each other's presence, but before the bar of him who is Eternal![19]

It is hard to belief that the Mesoamericanists continue to discount such incredibly clear descriptions as these for so many decades. But they do. Is that rational? It doesn't seem so. The strategy is evasion of and discounting of good, plain and clear evidence, which apologists of all sorts are so apt to do, when that kind of evidence gets in the way of a good pet theory. They should stop the tactic of trying to deny that Joseph Smith believed the hill in New York was the very hill where the Nephites were destroyed, and the records repository. Instead, they should just argue that they believe Joseph Smith was wrong on geography, and that they disagree with him. Because the historical record is pretty clear that he believed the Cumorah of the Nephite destruction was in New York. Joseph Fielding Smith stated:

> It must be conceded that this description fits perfectly the land of Cumorah in New York, . . . for the hill is in the proxim-

---

[19]. *Latter-day Saints Messenger and Advocate*, Vol. 2 No. 1, pp. 34-37 (Letter 7), emphasis added, spelling corrected

ity of the Great Lakes and also in the land of many rivers and fountains. . . . Further, the fact that all of his associates from the beginning down have spoken of it as the identical hill where Mormon and Moroni hid the records, must carry some weight. It is difficult for a reasonable person to believe that such men . . . [would] not be corrected by the Prophet, if that were not the fact. . . .

The first reference of this kind is found in the Messenger and Advocate, a paper published by the Church in 1834–5. In [it] . . . Oliver Cowdery . . . makes reference to this particular spot. . . .

The quibbler might say that this statement from Oliver Cowdery is merely the opinion of Oliver Cowdery and not the expression of the Prophet Joseph Smith. It should be remembered that these letters in which these statements are made were written at the Prophet's request and under his personal supervision. Surely, under these circumstances, he would not have permitted an error of this kind to creep into the record without correction.

At the commencement of these historical letters is found the following:

'That our narrative may be correct, and particularly the introduction, it is proper to inform our patrons, that our Brother J. Smith Jr., has offered to assist us. Indeed, there are many items connected with the fore part of this subject that render his labor indispensable. With his labor and with authentic documents now in our possession, we hope to render this a pleasing and agreeable narrative, well worth the examination and perusal of the saints.'

Later, during the Nauvoo period of the Church, and again under the direction of the Prophet Joseph Smith, these same letters by Oliver Cowdery, were published in the *Times and Sea-*

sons, without any thought of correction had this description of the Hill Cumorah been an error.[20]

Indeed, the Mesoamericanist arguments against this overwhelming and clear historical evidence amounts to mere quibbles. And they discount this clear logic from Joseph Fielding Smith. Though I strongly disagree with President Smith on other issues, this logic is clear, concise and correct. Orson Pratt stated:

> The Hill Cumorah is situated in western New York . . . the grand depository of all the numerous records of the ancient nations of the western continent was located in another department of the hill, and its contents under the charge of holy angels, until the day should come for them to be transferred to the sacred temple of Zion.[21]

While it is true that in my original book with Wayne May, when I was a proponent of what came to be called the Heartland Theory after that book was published, I tried to be somewhat comprehensive with these types of accounts. It isn't really necessary here, because all of the accounts about the repository of records in the cave are all well documented in Heartlander circles and in that original book. It's not my purpose here to be comprehensive this time around. But let me end this section with what I quoted from the Apocrypha in that book. In the book of 2 Baruch, it gives an account of when the Jerusalem temple was to be destroyed, to illustrate the very important principle of records being moved around at will by angels through the earth, that I alluded to in previous chapters:

> And another angel came down from heaven and said to

---

20. Smith, Joseph Fielding, 1954, *Doctrines of Salvation*, Vol 3, Salt Lake City, UT: Bookcraft, pp. 232-243

21. Lundwall, N. B., Temples of the Most High, Salt Lake City, Utah: Bookcraft, 1941, p. 232

them [i.e. the other angels with him], Hold your torches and do not light them before I say it to you. Because I was sent first to speak a word to the earth and then to deposit in it what the Lord, the Most High, has commanded me. And I saw that he descended in the Holy of Holies and that he took from there the veil, the holy ephod, the mercy seat, the two tables, the holy raiment of the priests, the altar of incense, the forty-eight precious stones with which the priests were clothed, and all the holy vessels of the tabernacle. And he said to the earth with a loud voice: Earth, earth, earth, hear the word of the mighty God, and receive the things which I commit to you, and guard them until the last times, so that you may restore them when you are ordered, so that strangers may not get possession of them. For the time has arrived when Jerusalem will also be delivered up for a time, until the moment that it will be said that it will be restored forever. And the earth opened its mouth and swallowed them up.[22]

This principle, I believe, is one of those things from the Apocrypha that ought to be taken seriously, backing up the point that the records can be moved, even if intrepid people actually do find or actually have found the cave. Whenever someone comes into such a place, it will be found empty. This same principle may extend to other evidence that has been withheld in terms of smoking-gun proof of the Book of Mormon:

And behold, if a man hide up a treasure in the earth, and the Lord shall say—Let it be accursed, because of the iniquity of him who hath hid it up—behold, it shall be accursed. And if the Lord shall say—Be thou accursed, that no man shall find thee from this time henceforth and forever—behold, no man getteth it henceforth and forever. (Helaman 12:18–19)

---

22. 2 Baruch 6:5-8

I'm certainly not extending this to all evidence for the Book of Mormon in terms of is archaeology and geography at all. I do not make a comprehensive statement that all evidence has been hidden by the Lord. The Lord has left us enough to come to some conclusions. And science has been allowed to find a lot. But for anyone to deny that this principle here of the hiding of evidences in the earth to say that it is not a factor at play that we need to recognize would be denying the word of the Lord here in this text. I never make excuses in general for what has not been found yet in archaeology or in science in general. I only recognize reality and live with it. I don't try to turn horses into tapirs the way some apologists do. But what I do say is that there are some facts that just have to be accepted as a fact of life, and not make excuses for it, in terms of certain things that just haven't been found yet. We have to acknowledge that we just don't have evidence for some things yet, and it is silly in a way to transform things into things that they just are not. Steel swords are not obsidian swords, etc. But something in the text of the Book of Mormon being mentioned in the text at all is prophetic insomuch that the evidence will definitely appear for such things someday, and vindicate it, if not at this time. The story of Heber J. Grant is a classic when he was confronted by a person that claimed that the Book of Mormon is false because of lack of evidence for cement. President Grant said that he didn't care whether it was in the time of his grandchildren or other descendants, but that the evidence for cement would appear. It wasn't long, and lo and behold, we find very high-quality cement throughout Mesoamerica. Teotihuacan, for example, is just full of the stuff. So, it's not for us to worry about such things. We only need to be patient until the Lord sees fit to prove his work in his own timing. It's up to us to keep our covenants and do the best we can with what he has given us in terms of partial evidences. What has not been found yet is never an excuse to ignore the piles of evidence that have already been found. But Anti-LDS people are fond of doing that to try to cast doubt, so that those that might be temporarily weak among us will be easy prey to them. They lie in wait to deceive.

But Mesoamericanist denial of just how strong the evidence actually is, especially the historical evidence, for Cumorah in New York as the actual place of the destruction of the Nephites isn't doing any favors for anyone. It is a misplaced apologetic tactic. Objective observers without a stake in the matter know what they see when they look at the historical evidence.

Mesoamericanists don't seem to realize that they can have Mesoamerica for the Land Southward, pair it with the Church tradition of Cumorah, and lose nothing whatsoever. And the fact that the Dual Heartland Theory transforms the Hopewell into the Northern Urban Heartland, a satellite of the Maya civilization, is nothing short of amazing and actually ought to be welcomed. They trumpet how advanced the Maya were but deny that the Northern transplants of that civilization actually did transfer their technologies and their way of life, their culture and literacy, to the North, and they absorbed the natives of the area into their nation. Nothing is lost whatsoever by these facts. This is not a denial of any of the strong evidences from the archaeology from the Hopewell/Adena that they were an advanced and populous civilization.

The only things that are ought not to be considered evidences and ought to be rejected from Book of Mormon Archaeology are things which don't help our case that simply aren't real.

The association I had with artifacts of a controversial nature two decades ago began with my involvement with the book, *This Land: Zarahemla and the Nephite Nation*. I was skeptical of artifacts like the Burrows Cave and Michigan "relics," but co-author Wayne May was a believer in them. I entered into a compromise as a business arrangement, where I took a neutral stance on these controversial artifacts in the book, with a "wait and see" approach in order to find common ground with the co-author. It was an unfortunate episode, and I have learned a lot from it. After the book was published, I found out some "artifacts" from these so-called "collections" had been deemed fraudulent by scientific testing, but only *after* the book had already been published. I ended up retracting the book, and the business arrangement with the co-author fell apart as a result. It isn't only because of this sad occurrence that I am saying what I am saying here. I should have been wiser.

I'm bringing this up as a cautionary tale not only because people wondered what happened to me and why I suddenly disappeared from the Heartlander scene. It wasn't only because later I changed my mind on the geographical placements, as you can see. But I am urging people that they ought to strive to be fully informed about what they are dealing with. The use of artifacts that science has deemed fraudulent in any form or fashion as a defense for the Book of Mormon hurts our case and makes people not take us seriously. We'll all be better off as a Church if we entirely divest ourselves of them.

Chapter 11

Internal Textual Evidence for a Limited, Tropical Geography, at Least for the Land Southward

The Book of Mormon text itself provides numerous internal clues that strongly suggest a geographically confined and tropical setting, laying the groundwork for the Mesoamerican model. Nevertheless, as we have seen in this book, to say that the Land of Desolation was jammed in there with that confined area is simply not true. But as for the Land Southward, the narrative consistently describes travel times that imply a relatively small geographical area, with journeys often taking "days or less." This internal consistency regarding distances is a foundational argument for a limited geography, as it clearly excludes a hemispheric model.

Furthermore, the text provides strong indicators of a tropical or semi-tropical climate. There is no indication of cold or snow given in the text, while heat is explicitly mentioned. Descriptions of climate suggest warm and mild conditions were typical (Alma 51:33), with only a single, atypical reference to hail and no mention of snow and ice in the land of promise. This consistent absence of cold-weather phenomena, combined with the explicit presence of tropical resources, directly challenges a Hemispheric Model and the Heartland Model. Neither that would necessarily include vast temperate or arctic regions of North and South America. The lack of certain expected environmental features becomes as powerful an argument against a broad geography as the presence of others is for a limited, tropical one. This illustrates how internal textual details, sometimes even by their omissions, can serve as powerful constraints, effectively pointing away from the traditional Hemispherical model.

The narrative also depicts densely populated nations with millions of inhabitants, extensive commerce, widespread literacy, metallurgy, and finely woven textiles. Such a high level of societal complexity and population density, capable of supporting large-scale warfare and city-building, is far more consistent with known ancient Mesoamerican civilizations than with the area of the United States, where things were still very advanced and shared a similar culture, but still more spread out with much land to spare. The Book of Mormon mentions the abundance of precious metals like gold and silver (1 Ne. 18:25; Hel. 6:9; Ether 9:17; 10:23), both of which are historically plentiful in Mesoamerica, aligning with these textual descriptions. Additionally, "fine pearls" are noted as an important luxury item (4 Ne. 1:24). The most precious pearls originate from tropical to subtropical seas and were abundant off the coasts of southern Mexico, prized by Mesoamerican peoples from preclassic times. The catastrophic destruction described in 3 Nephi 8–10, including earthquakes and darkness, are consistent with volcanic events. Middle America is renowned as one of the most volcanically active regions in the world, providing a geological plausibility for these textual events within a Mesoamerican context.

The "land southward," where the Lehites initially landed around 589 BC, is described as being "nearly surrounded by seas." Similarly, the Jaredite land, later known as the "Land Northward," was described as being surrounded by four "seas" with a feature called a "Narrow Neck" serving as a link to the "Land Southward," which they generally avoided except for hunting. This singular, defining feature acts as the crucial hinge point that dictates the scale and orientation of the entire Book of Mormon geography. Its specific identification in this book (the Chivela Pass) directly determines the overall size and placement of the lands. The shift from Panama to the Tehuantepec area was a critical step in confining the geography to Mesoamerica, drastically reducing the scale, although other theorists equate the Isthmus of Tehuantepec as the neck, but many have criticized it as too broad. I share that skepticism of this being a description of the isthmus itself, which is why I identify the wording of "neck" as being instead synonymous with "pass" or "passage." The "narrow neck" is not merely a descriptive detail but the linchpin of any proposed geography. Its specific characteristics (linking best with the Chivela Pass), in conjunction with the Olmec heartland being right by it, and also a fabulous candidate for the place where the sea divides the land being the series of lagoons on the coast, become the most powerful textual filter for plausible real-world locations.

The Book of Mormon explicitly mentions an "east sea" and a "west sea." In a Mesoamerican context, these are typically identified as the Pacific Ocean and the Atlantic Ocean (or its regional components, such as the Gulf of Mexico or the Caribbean Sea). The concept of "four seas" (sea south, sea north, sea west, and sea east) is mentioned in Helaman 3:8. While the Pacific and Atlantic easily account for the east and west seas, the identification of distinct "north" and "south" seas within Mesoamerica presents an interpretive challenge for some models, but being that my model and Lance Weaver's model are examples of Continental-spanning geographies with an extended Land Northward that presents less of a problem for our category of geography. Some Mesoamericanists, like Richard Hauck, have attempted to literally map all four seas, while others, including John Sorenson and John Clark, suggest a non-literal or metaphorical interpretation for the north and south seas, arguing that a literal interpretation would require an unlikely land orientation or is simply unfeasible. The ongoing discussion about the "four seas" further underscores how even seemingly minor textual details around this central feature can lead to significant interpretive challenges and shape the precise contours of the proposed geography. The River Sidon is another prominent hydrological feature in the "land southward." While John Sorenson identifies the Grijalva River as the River Sidon, other scholars within the Mesoamerican framework, including me, strongly advocate for the Usumacinta River. This ongoing debate within the Mesoamerican model itself highlights the specificity and complexity involved in correlating textual descriptions with real-world geographical features.

The following table summarizes these key geographical descriptors and their proposed Mesoamerican correlates:

Book of Mormon Descriptor	Possible Mesoamerican Correlate
Narrow Neck of Land	Isthmus of Tehuantepec, Chivela Pass or Coastal Corridor
Land Southward (nearly surrounded by seas)	Southern Guatemala/Mexican states of Chiapas & Tabasco
Land Northward	Southern Mexico/Oaxaca & Southern Veracruz (or also in the case of the Continental-spanning models, to include the Hopewell/Adena region of Eastern North America)
East Sea	Pacific Ocean
West Sea	Gulf of Mexico/Caribbean Sea
North Sea (Helaman 3:8)	Interpretive challenge, some propose Gulf of Mexico for north, but for a Continental-spanning Geography it is either the Arctic Ocean or perhaps the Hudson Bay

Internal Textual Evidence for a Limited, Tropical Geography

Book of Mormon Descriptor	Possible Mesoamerican Correlate
South Sea (Helaman 3:8)	Interpretive challenge, some propose Pacific for south
River Sidon	Grijalva River / Usumacinta River
Climate (no snow, heat, warm/mild)	Tropical/Semi-tropical climate
Resources (gold, silver, fine pearls)	Abundant gold/silver/pearls in region
Geological Events (volcanic activity, earthquakes, darkness)	Volcanically active region
Travel Times (short, days/less)	Distances consistent with Mesoamerica, at least for the Land Southward
Population Density (millions, large cities)	Archaeological evidence of dense populations and large cities

Chapter 12

John L. Sorenson's Foundational Mesoamerican Model

John L. Sorenson is widely recognized as a pivotal figure in Book of Mormon geography, with his name becoming "synonymous with a specific geographic correlation between the Book of Mormon and a Mesoamerican geography." His landmark publication, *An Ancient American Setting for the Book of Mormon* (1985), was the culmination of decades of meticulous research.

Sorenson's methodology is characterized by a rigorous, interdisciplinary approach that moves beyond simple mapping. He meticulously compares the entire spectrum of cultural and historical information found within the Book of Mormon text with data derived from reliable archaeological and anthropological studies of Mesoamerica. His process is marked by asking numerous carefully weighed questions rather than providing immediate answers. These questions include fundamental inquiries such as "Who were these people?," "What might they have looked like?," "Who were their neighbors?," "How many of them were there?," and "How did they live, eat, speak, work, or fight?" By systematically addressing these questions and matching Book of Mormon data with Mesoamerican studies, he aimed to develop a "picture or model of how Book of Mormon events took place" that is plausible and internally consistent, although I disagree of course on the extent of the Land Northward and the Grijalva as the Sidon. But the impact of his work cannot be underestimated.

Crucially, Sorenson explicitly stated that his work was not intended to "test" the Book of Mormon for its truthfulness or provide "sure proof." He acknowledged that conclusive results can never be obtained that way. This represents a significant methodological and epistemological maturation in Book of Mormon scholarship. Instead of seeking irrefutable evidence, which is often unattainable in historical archaeology, his approach focused on establishing a "credible model" and a "plausible geographical, historical, and cultural setting." This allows him to focus on demonstrating how the events could have happened within a complex, real-world context without somehow proving it. This approach also helps manage expectations among believers.

A cornerstone of Sorenson's model is his belief for the Isthmus of Tehuantepec as the 'narrow neck of land'". This identification is central to his limited geography.

Sorenson's contribution extended beyond mere physical geography into the realm of culture and history. His correlations between the Book of Mormon narrative and ancient Mesoamerican societies were impressive. His later publication, *Mormon's Codex*, further built up these arguments.

The Book of Mormon Archaeological Forum (BMAF), before it was swallowed up by what became Scripture Central, was an organization focused on Mesoamerican research. Activists associated with it like Joe V. Andersen and Ted Dee Stoddard directly challenged the Heartland theory with a lot of energy.

Many other Mesoamericanist activists over time have contributed greatly as well to the corpus of information brought forth that has made Mesoamerica the most plausible candidate for the Land Southward in my mind.

Chapter 13

Conclusion: Converging Arguments for a Mesoamerican Land Southward and a North American Land Northward

In summary, the scholarly consensus for confining the Book of Mormon's "land southward" to Mesoamerica is built upon a compelling convergence of arguments derived from rigorous internal textual analysis and external archaeological and anthropological correlations. The field of Book of Mormon geography has undergone a significant maturation. It began with broad, sometimes speculative, hemispheric models, often driven by a general understanding of the continents. John Sorenson then introduced a systematic, detailed, and interdisciplinary approach, shifting the focus from simply "proving" the book to establishing its "plausibility" within a real-world setting, grounded in extensive internal textual analysis and external archaeological data. This provided a much-needed academic framework. The Book of Mormon Archaeological Forum (BMAF) fostered a community of scholars, generating extensive research, publishing detailed analyses, and engaging in robust debates, further solidifying the Mesoamerican model and emphasizing the importance of geographical constraints. Finally, John Clark brought a crucial layer of archaeological rigor and epistemological caution, highlighting the limits of scientific "proof" for religious claims while affirming the corroborative value of archaeological ev-

Conclusion: Converging Arguments

idence. The combined efforts and intellectual evolution represented by these scholars and organizations have led to a more sophisticated, nuanced, and academically grounded discourse on Book of Mormon geography.

The most fundamental and consistently compelling argument for a limited geography, and by extension, for Mesoamerica as the "land southward," remains the internal textual evidence itself. The Book of Mormon's detailed descriptions of limited travel times, the necessity of a specific "narrow neck of land," and its consistent portrayal of a tropical environment with specific resources and geological phenomena, collectively render hemispheric models untenable. These internal constraints are the primary drivers that point to a geographically restricted and climatically specific region.

John L. Sorenson's pioneering work systematically correlated hundreds of geographical, historical, and cultural details from the Book of Mormon with Mesoamerican data.

With Wayne May and myself being the early pioneers of the Heartland model (not to forget the prior important contributions of those that even came before us in less systematized North American Geographies like Duane Erickson, Delbert Curtis, etc.), we sowed the seeds of a strong and vigorous movement. With the continued important contributions of others like Rod Meldrum, Jonathan Neville and many others, the movement continues to bring forth important research. And it has been extremely important in the effort to push back on the Mesoamericanists on the issue of Cumorah.

Nevertheless, the divide between Heartlanders and Mesoamericanists needs to be healed. I hope to continue challenging the Mesoamericanists to get them to acknowledge the plausibility, once again, of the New York Cumorah and to recognize how significant the Urban Society of the North is, and to stop minimizing it. I want them to stop acting as if they are experts in North American Archaeology when it is not their specialty. I want them to get truly educated about it.

I seek to challenge my fellow brothers that believe in the New York Cumorah, the Heartlanders, to reconsider their candidate for their Land Southward, and take very seriously the reasons that led me to abandon the Heartland Theory while maintaining a belief in a New York Cumorah and an important Northern extension of Book of Mormon Urban Society in the North, but recognizing what is really in the South, and how they are intimately connected.

I intend to turn the tide and unite the Heartland region of North America with the Heartland of the South in Mesoamerica, and help all of these good people see the big picture, so it is all circumscribed into one great whole, to unify them. I predict that Continent-spanning geographic models like mine and Lance Weaver's will be the future of Book of Mormon Geography for decades to come, and will become more popular as time goes by, when proponents of both leading models will see the wisdom in unification, and putting aside differences, in order to martial the strengths of both Mesoamerica and the Great Lakes Region.

I was a pioneer in the Heartland Model, and I left it behind. It was but a stage in the evolution of my thinking. I am now a pioneer in this new push, aiming to unify both sides, to eventually end any residual contention between models and theorists and bring us to a "unity of the faith" so to speak, at least with Geography, so there is less contention in the Church as a whole.

www.ingramcontent.com/pod-product-compliance
Lightning Source LLC
Chambersburg PA
CBHW080508110426
42742CB00017B/3030